Springer
Berlin
Heidelberg
New York
Barcelona
Hong Kong
London
Milan
Paris
Singapore
Tokyo

A. Jensen
A. la Cour-Harbo

Ripples in
Mathematics
The Discrete Wavelet
Transform

Springer

Arne Jensen
Aalborg University
Department of Mathematical Sciences
Fredrik Bajers Vej 7
9220 Aalborg, Denmark
e-mail: matarne@math.auc.dk

Anders la Cour-Harbo
Aalborg University
Department of Control Engineering
Fredrik Bajers Vej 7C
9220 Aalborg, Denmark
e-mail: alc@control.auc.dk

Library of Congress Cataloging-in-Publication Data

Jensen, A. (Arne), 1950-
    Ripples in mathematics : the discrete wavelet transform / A. Jensen, A. La Cour-Harbo.
    p. cm.
    Includes bibliographical references and index.
    ISBN 3540416625 (softcover : alk. paper)
    1. Wavelets (Mathematics) I. La Cour-Harbo, A. (Anders), 1973- II. Title.

QA403.3 .J46 2001
515'.2433--dc21

2001020907

ISBN 3-540-41662-5 Springer-Verlag Berlin Heidelberg New York

Mathematics Subject Classification (2000): 42-01, 42C40, 65-01, 65T60, 94-01, 94A12

MATLAB® is a registred trademark of The MathWorks, Inc.

Springer-Verlag Berlin Heidelberg New York
a member of BertelsmannSpringer Science+Business Media GmbH

http://www.springer.de

© Springer-Verlag Berlin Heidelberg 2001
Printed in Germany

Cover design: *Künkel & Lopka,* Heidelberg
Typesetting by the authors using a LaTeXmacro package
Printed on acid-free paper   SPIN 12071593   46/3180ck-5 4 3

Preface

Yet another book on wavelets. There are many books on wavelets available, written for readers with different backgrounds. But the topic is becoming ever more important in mainstream signal processing, since the new JPEG2000 standard is based on wavelet techniques. Wavelet techniques are also important in the MPEG-4 standard.

So we thought that there might be room for yet another book on wavelets. This one is limited in scope, since it only covers the discrete wavelet transform, which is central in modern digital signal processing. The presentation is based on the lifting technique discovered by W. Sweldens in 1994. Due to a result by I. Daubechies and W. Sweldens from 1996 this approach covers the same class of discrete wavelet transforms as the one based on two channel filter banks with perfect reconstruction.

The goal of this book is to enable readers, with modest backgrounds in mathematics, signal analysis, and programming, to understand wavelet based techniques in signal analysis, and perhaps to enable them to apply such methods to real world problems.

The book started as a set of lecture notes, written in Danish, for a group of teachers of signal analysis at Danish Engineering Colleges. The material has also been presented to groups of engineers working in industry, and used in mathematics courses at Aalborg University.

We would like to acknowledge the influence of the work by W. Sweldens [25, 26] on this book. Without his lifting idea we would not have been able to write this book. We would also like to acknowledge the influence of the paper [20] by C. Mulcahy. His idea of introducing the wavelet transform using a signal with 8 samples appealed very much to us, so we have used it in Chap. 2 to introduce the wavelet transform, and many times later to give simple examples illustrating the general ideas. It is surprising how much of wavelet theory one can explain using such simple examples.

This book is an exposition of existing, and in many cases well-known, results on wavelet theory. For this reason we have not provided detailed references to the contributions of the many authors working in this area. We acknowledge all their contributions, but defer to the textbooks mentioned in the last chapter for detailed references.

Tokyo, December 2000   *Arne Jensen*

Aalborg, December 2000   *Anders la Cour-Harbo*

Contents

1. Introduction

This book gives an introduction to the discrete wavelet transform, and to some of its generalizations. The transforms are defined and interpreted. Some examples of applications are given, and the implementation on the computer is described in detail. The book is limited to the discrete wavelet transform, which means that the continuous version of the wavelet transform is not presented at all. One of the reasons for this choice is the intention that the book should be accessible to readers with rather modest mathematical prerequisites. Another reason is that for readers with good mathematical prerequisites there exists a large number of excellent books presenting the continuous (and often also the discrete) versions of the wavelet transform.

The book is written for at least three different audiences. (i) Students of electrical engineering that need a background in wavelets in order to understand the current standards in the field. (ii) Electrical engineers working in industry that need to get some background in wavelets in order to apply these to their own problems in signal processing. (iii) Undergraduate mathematics students that want to see the power and applicability of modern mathematics in signal processing.

In this introduction we first describe the prerequisites, then we give a short guide to the book, and finally we give some background information.

1.1 Prerequisites

The prerequisites for reading this book are quite modest, at least for the first six chapters. For these chapters familiarity with calculus and linear algebra will suffice. The numerous American undergraduate texts on calculus and linear algebra contain more material than is needed. From Chap. 7 onwards we assume familiarity with either the basic concepts in digital signal processing, as presented in for example [22, 23] (or any introductory text on digital signal processing), or with Fourier series. What is needed is the Fourier series, and the z-transform formulation of Fourier series, together with basic concepts from filter theory, or, in mathematical terms, elementary results on convolution of sequences. This chapter is somewhat more difficult to read than the previous chapters, but the material is essential for a real understanding of the wavelet transforms.

The ultimate goal of this book is to enable the reader to use the discrete wavelet transform on real world problems. For this goal to be realized it is necessary that the reader carries out experiments on the computer. We have chosen MATLAB as the environment for computations, since it is particularly well suited to signal processing. We give many examples and exercises using MATLAB. A few examples are also given using the C language, but these are entirely optional. The MATLAB environment is easy to use, so a modest background in programming will suffice. In Chap. 13 we provide a number of examples of applications of the various wavelet transforms, based on a public domain toolbox, so no programming skills are needed to go through the examples in that chapter.

1.2 Guide to the Book

The reader should first go through Chap. 2 to Chap. 6 without solving the computer exercises, and then go through the first part of Chap. 13. After that the reader should return to the first chapters and do the computer exercises. The first part of the book is based on the so-called lifting technique, which gives a very easy introduction to the discrete wavelet transform. For the reader with some previous knowledge of the wavelet transform we give some background information on the lifting technique in the next section.

In Chap. 7 we establish the connection between the lifting technique and the more usual filter bank approach to the wavelet transform. The proof and the detailed discussion of the main result is postponed to Chap. 12.

In Chap. 8 we define the generalization of the wavelet transform called wavelet packets. This leads to a very large number of possible representations of a given signal, but fortunately there is a fast search algorithm associated with wavelet packets. In Chap. 9 we interpret the transforms in time and frequency, and for this purpose we introduce the time-frequency plane. One should note that the interpretation of wavelet packet transforms is not easy. Computer experiments can help the reader to understand the properties of this class of transforms. The rather complicated behavior with respect to time and frequency is on the other hand one of the reasons why wavelets and wavelet packets have been so successful in applications to data compression and denoising of signals.

Up to this point we have not dealt with an essential problem in the theory, and in particular in the applications. Everything presented in the previous chapters works without problems, when applied to infinitely long signals. But in the real world we always deal with finite length signals. There are problems at the beginning, and at the end, of a finite signal, when one wants to carry out a wavelet analysis of such a signal. We refer to this as the boundary problem. In Chap. 10 we present several solutions to this boundary problem. There is no universal solution. One has to choose a boundary correction method adapted to the class of signals under consideration.

In Chap. 11 we show in detail how to implement wavelet transforms and wavelet packet transforms in the MATLAB environment. Several different approaches are discussed. Some examples of C code implementations are also given. These are optional. In Chap. 12 we complete the results in Chap. 7 on filters and lifting steps.

In Chap. 13 we use MATLAB to demonstrate some of the capabilities of wavelets applied to various signals. This chapter is based on the public domain toolbox called *Uvi_Wave*. At this point the reader should begin to appreciate the advantages of the wavelet transforms in dealing with signals with transients. After this chapter the reader should review the previous chapters and do further experiments on the computer.

The last chapter contains an overview of some applications of wavelets. We have chosen not to give detailed presentations, since each application has specific prerequisites, quite different from those assumed in the preceding chapters. Instead we give references to the literature, and to web sites with relevant information. The chapter also contains suggestions on how to learn more about wavelets. This book covers only a small part of the by now huge wavelet theory. There are a few appendices containing supplementary material. Some additional material, and the relevant MATLAB M-files, are available electronically, at the URL

`http://www.bigfoot.com/~alch/ripples.html`

Finally, at the end of the book we give some references. There are references to a few of the numerous books on wavelets and to some research papers. The latter are included in order to acknowledge some sources. They are probably inaccessible to most readers of this book.

1.3 Background Information

In this section we assume that the reader has some familiarity with the usual presentations of wavelet theory, as for example given in [5]. Readers without this background should go directly to Chap. 2.

We will here try to explain how our approach differs from the most common ones in the current wavelet literature. We will do this by sketching the development of wavelets. Our description is very short and incomplete. A good description of the history of wavelets is given in [13]. Wavelet analysis started with the work by A. Grossmann and J. Morlet in the beginning of the eighties. J. Morlet, working for a French oil company, devised a method for analyzing transient seismic signals, based on an analogy with the windowed Fourier transform (Gabor analysis). He replaced the window function by a function ψ, well localized in time and frequency (for example a Gaussian), and replaced translation in frequency by scaling. The transform is defined as

$$CWT(f; a, b) = \int_{-\infty}^{\infty} f(t) a^{-1/2} \overline{\psi}(a^{-1}(t - b)) dt .$$

Under some additional conditions on ψ the transform is invertible. This transform turned out the be better than Fourier analysis in handling transient signals. The two authors gave the name 'ondelette', in English 'wavelet,' to the analyzing function ψ. Connections to quantum mechanics were also established in the early papers.

It turned out that this continuous wavelet transform was not that easy to apply. In 1985 Y. Meyer discovered that by using certain discrete values of the two parameters a, b, one could get an orthonormal basis for the Hilbert space $L^2(\mathbf{R})$. More precisely, the basis is of the form

$$\{2^{j/2}\psi(2^j t - k)\}_{j,k\in\mathbf{Z}} \ .$$

The first constructions of such ψ were difficult.

The underlying mathematical structure was discovered by S. Mallat and Y. Meyer in 1987. This structure is called a multiresolution analysis. Combining ideas from Fourier analysis with ideas from signal processing (two channel filter banks) and vision (pyramidal algorithms) this leads to a characterization of functions ψ, which generate a wavelet basis. At the same time this framework establishes a close connection between wavelets and two channel filter banks with perfect reconstruction. Another result obtained by S. Mallat was a fast algorithm for the computation of the coefficients for certain decompositions in a wavelet basis.

In 1988 I. Daubechies used the connection with filter theory to construct a family of wavelets, with compact support, and with differentiability to a prescribed finite order. Infinitely often differentiable wavelets with compact support do not exist.

From this point onwards the wavelet theory and its applications underwent a very fast development. We will only mention one important event. In a paper [25], which appeared as a preprint in 1994, W. Sweldens introduced a method called the 'lifting technique,' which allowed one to improve properties of existing wavelet transforms. I. Daubechies and W. Sweldens [7] proved that all finite filters related to wavelets can be obtained using the lifting technique. The lifting technique has many advantages, and it is now part of mainstream signal analysis. For example, the new JPEG2000 standard is based on the lifting technique, and the lifting technique is also part of the MPEG-4 standard.

The main impact of the result by I. Daubechies and W. Sweldens, in relation to this book, is that one can start from the lifting technique and use it to give a direct and simple definition of the discrete wavelet transform. This is precisely what we have done, and this is how our approach differs from the more usual ones.

If one wants to go on and study the wavelet bases in $L^2(\mathbf{R})$, then one faces the problem that not all discrete wavelet transforms lead to bases. But there are two complete characterization available, one in the work by A. Cohen, see for example [4], and a different one in the work by W. Lawton, see for example [5, 15]. From this point onwards the mathematics becomes highly nontrivial. We choose to stop our exposition here. The reader will have to have the necessary mathematical background to continue, and with that background there is a large number of excellent books with which to continue.

2. A First Example

In this chapter we introduce the discrete wavelet transform, often referred to as DWT, through a simple example, which will reveal some of its essential features. This idea is due to C. Mulcahy [20], and we use his example, with a minor modification.

2.1 The Example

The first example is very simple. We take a digital signal consisting of just 8 samples,

$$56, \quad 40, \quad 8, \quad 24, \quad 48, \quad 48, \quad 40, \quad 16 \,.$$

We display these numbers in the first row of Table 2.1. We assume that these numbers are not random, but contain some structures that we want to extract. We could for example assume that there is some correlation between a number and its immediate successor, so we take the numbers in pairs and compute the mean, and the difference between the first member of the pair and the computed mean. The second row contains the four means followed by the four differences, the latter being typeset in boldface. We then leave the four differences unchanged and apply the mean and difference computations to the first four entries. We repeat this procedure once more. The fourth row then contains a first entry, which is the mean of the original 8 numbers, and the 7 calculated differences. The boldface entries in the table are here called the details of the signal.

Table 2.1. Mean and difference computation. Differences are in boldface type

56	40	8	24	48	48	40	16
48	16	48	28	8	−8	0	12
32	38	16	10	8	−8	0	12
35	−3	16	10	8	−8	0	12

It is important to observe that no information has been lost in this transformation of the first row into the fourth row. This means that we can reverse the calculation. Beginning with the last row, we compute the first two entries in the third row as $32 = 35 + (-3)$ and $38 = 35 - (-3)$, Analogously, the first 4 entries in the second row are calculated as $48 = 32 + (16)$, $16 = 32 - (16)$, $48 = 38 + (10)$, and finally $28 = 38 - (10)$. Repeating this procedure we get the first row in the table.

Do we gain anything from this change of representation of the signal? In other words, does the signal in the fourth row exhibit some nice features not seen in the original signal? One thing is immediately evident. The numbers in the fourth row are generally smaller than the original numbers. So we have achieved some kind of loss-free compression by reducing the dynamic range of the signal. By loss-free we mean that we can transform back to the original signal, without any loss of information. We could measure the dynamics of the signal by counting the number of digits used to represent it. The first row contains 15 digits. The last row contains 12 digits and two negative signs. So in this example the compression is not very large. But it is easy to give other examples, where the compression of the dynamic range can be substantial.

We see in this example the pair 48, 48, where the difference of course is zero. Suppose that after transformation we find that many difference entries are zero. Then we can store the transformed signal more efficiently by only storing the non-zero entries (and their locations).

Let us now suppose that we are willing to accept a certain loss of quality in the signal, if we can get a higher rate of compression. We can try to process our signal, or better, our transformed signal. One technique is called *thresholding*. We choose a threshold and decide to put all entries with an absolute value less than this threshold equal to zero. Let us in our example choose 4 as the threshold. This means that we in Table 2.1 replace the entry **−3** by **0** and then perform the reconstruction. The result is in Table 2.2.

Table 2.2. Reconstruction with threshold 4

59	43	11	27	45	45	37	13
51	19	45	25	8	−8	0	12
35	35	16	10	8	−8	0	12
35	0	16	10	8	−8	0	12

The original and transformed signal are both shown in Fig. 2.1. We have chosen to join the given points by straight line segments to get a good visualization of the signals. Clearly the two graphs differ very little. If presented in separate plots, it would be difficult to tell them apart. Now let us perform a more drastic compression. This time we choose the threshold equal to 9. The computations are given in Table 2.3, and the graphs are plotted in Fig. 2.2. Notice that the peaks in the original signal have been flattened. We also note that the signal now is represented by only four non-zero entries.

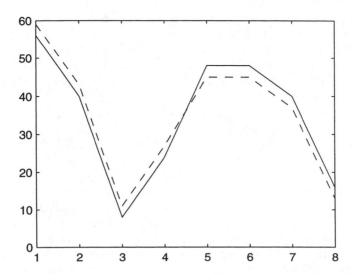

Fig. 2.1. Original signal and modified signal (*dashed line*) with threshold 4

Table 2.3. Reconstruction with threshold 9

51	51	19	19	45	45	37	13
51	19	45	25	0	0	0	12
35	35	**16**	**10**	0	0	0	**12**
35	**0**	**16**	**10**	0	0	0	**12**

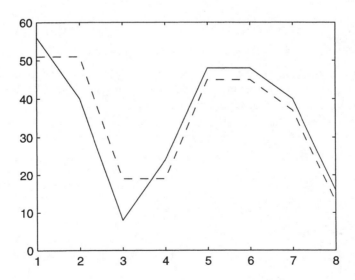

Fig. 2.2. Original signal and modified signal (*dashed line*) with threshold 9

We note that there are several variations of the procedure used here. We could have stored averages and differences, or we could have used the difference between the second element of the pair and the computed average. The first choice will lead to boldface entries in the tables that can be obtained from the computed ones by multiplication by a factor -2. The second variant is obtained by multiplication by -1.

2.2 Generalizations

The above procedure can of course be performed on any signal of length 2^N, and will lead to a table with $N + 1$ rows, where the first row is the original signal. If the given signal has a length different from a power of 2, then we will have to do some additional operations on the signal to compensate for that. One possibility is to add samples with value zero to one or both ends of the signal until a length of 2^N is achieved. This is referred to as *zero padding*.

The transformation performed by successively calculating means and differences of a signal is an example of the discrete wavelet transform. It can be undone by simply reversing the steps performed. We have also seen that the transformed signal may reveal features not easily seen or detected in the original signal. All these phenomena are consequences of the properties of the discrete wavelet transform, as we will see in the following chapters.

Exercises

2.1 Verify the computations in the tables in this chapter.

2.2 Give some other examples, using for example signals of length 16.

2.3 Write some simple functions (in the programming language of your choice) to perform transformation of signals of length 256 or 512. With these functions perform some experiments with zero padding.

3. The Discrete Wavelet Transform via Lifting

We now introduce the discrete wavelet transform via a procedure, which is known as 'lifting' in the literature. This procedure was introduced by W. Sweldens in a series of papers [25, 26].

3.1 The First Example Again

We start by looking at the example from Sect. 2.1 again, and at the same time introduce some useful notation.

A discrete signal is an ordered (finite) sequence of real or complex numbers. Perhaps it has been obtained from a continuous signal by sampling, or the signal was from the beginning discrete. For example, if the signal is the result of playing back an audio CD. The number of samples is called the length of the signal. A finite signal of length N is represented as

$$x[1], x[2], x[3], \ldots, x[N] .$$

In some cases it is more convenient to start the enumeration with index zero

$$x[0], x[1], x[2], x[3], \ldots, x[N-1] .$$

We will use both conventions. The first enumeration scheme is used in MAT-LAB implementations. The second scheme is used in theoretical discussions.

In the theory presented here one can deal with signals of arbitrary length, including infinite length. When considering a signal of infinite length it is often convenient not to assume a starting point for the enumeration. This means that we will enumerate infinite signals using all integers, both positive and negative. For infinite length signals we will require that they have finite energy.

Finite signals can be considered a special case of the infinite signals, where we, for some integers $n_{\text{first}} < n_{\text{last}}$, know that $x[n] = 0$ for all n such that $n < n_{\text{first}}$, and also $x[n] = 0$ for all $n > n_{\text{last}}$. We use several different forms of notation for infinite signals. A signal can be denoted by a boldface letter like $\mathbf{x}$, or we can write $\{x[n]\}_{n \in \mathbf{Z}}$. We have chosen the notation $x[n]$ in favor of x_n, since it emphasizes that a signal is regarded as a function of the (discrete) parameter n. This also has the advantage that we can denote sequences by $\mathbf{x}_1$ and $\mathbf{x}_2$, or in the more detailed notation $\{x_1[n]\}_{n\in\mathbf{Z}}$ and $\{x_2[n]\}_{n\in\mathbf{Z}}$.

The condition for finite energy is for an infinity signal

$$\sum_{n=-\infty}^{\infty} |x[n]|^2 < \infty \,.$$

A finite signal always satisfies this condition. In the sequel the reader can assume that all signals are finite. Note that there are several technicalities involved in treating infinite sums, but since they are irrelevant for most of what we want to present here, we will omit these technicalities. The set of all signals with finite energy is denoted by $\ell^2(\mathbf{Z})$ in the literature. We will use this convenient notation below. Often we use the mathematical term sequence instead of . Sometimes we also use the term vector, in particular in connection with use of results from linear algebra.

Let us now return to the example in Sect. 2.1. We took a pair of numbers a, b and computed the mean, and the difference between the first entry and the mean

$$s = \frac{a+b}{2} \,, \tag{3.1}$$

$$d = a - s \,. \tag{3.2}$$

The inverse transform is then

$$a = s + d \,, \tag{3.3}$$

$$b = s - d \,. \tag{3.4}$$

As mentioned at the end of Sect. 2.1 we could have chosen another computation of the difference, as in

$$\mu = \frac{a+b}{2} \,, \tag{3.5}$$

$$\delta = b - a \,. \tag{3.6}$$

There is an important thing to be noticed here. When we talk about mean and difference of a pair of samples, as we have done in the previous chapter, the most obvious calculations are (3.5) and (3.6). And yet we have in Chap. 2 used (3.1) and (3.2) (the same sum, but a different difference). The reason for choosing this form is the following. Once s has been calculated in (3.1), the b is no longer needed, since it does not appear in (3.2) (this is in contrast to (3.6), where both a and b are needed). Thus in a computer the memory space used to store b can be used to store s. And once d in (3.2) has been calculated, we do not need a anymore. In the computer memory we can therefore also replace a with d.

First step: $a, b \rightarrow a, s$

Second step: $a, s \rightarrow d, s$

or with the operations indicated explicitly:

First step: $a, b \rightarrow a, \frac{1}{2}(a + b)$

Second step: $a, s \rightarrow a - s, s$.

Since we do not need extra memory to perform this transform, we refer to it as an *'in place' transform*. The inversion can also be performed 'in place,' namely as

First step: $d, s \rightarrow a, s$

Second step: $a, s \rightarrow a, b$

or with the operations given explicitly:

First step: $d, s \rightarrow d + s, s$

Second step: $a, s \rightarrow a, 2s - a$.

The difference between (3.2) and (3.5) might seem trivial and unimportant, but the replacement of old values with newly calculated ones is nonetheless one of the key features of the lifting scheme. One can see the importance, when one considers the memory space needed for transforming very long signals.

Actually, the computation in (3.5) and (3.6) can also be performed 'in place'. In this case we should start by computing the difference, as shown here

$$a, b \rightarrow a, \delta = b - a \rightarrow \mu = a + \delta/2, \delta . \tag{3.7}$$

Note that $\mu = a + \delta/2 = a + (b - a)/2 = (a + b)/2$ actually is the mean value. The inversion is performed as

$$\mu, \delta \rightarrow a = \mu - \delta/2, \delta \rightarrow a, b = a + \delta . \tag{3.8}$$

One important lesson to be learned from these computations is that essentially the *same* transform can have *different* implementations. In this example the differences are minor, but later we will see examples, where there can be more substantial differences.

3.2 Definition of Lifting

The transform that uses means and differences, brings us to the definition of the lifting operation. The two operations, mean and difference, can be viewed as special cases of more general operations. Remember that we previously (in the beginning of Chap. 2) assumed that there is some correlation between two successive samples, and we therefore computed the difference. If two samples are almost equal the difference is, of course, small, and it is therefore obvious to think of the first sample as a *prediction* of the second sample. It is a good prediction, if the difference is small. We can use other prediction steps than one based on just the previous sample. Examples are given later.

We also calculated the mean of the two samples. This can be viewed in two ways. Either as an operation, which preserves some properties of the original signal (later we shall see how the mean value (and sometimes also the energy) of a signal is preserved during transformation), or as an extraction of an essential features of the signal. The latter viewpoint is based on the fact that the pair-wise mean values contain the overall structure of the signal, but with only half the number of samples. We use the word *update* for the this operation. As with the prediction the update operation can be more sophisticated than just calculating the mean. An example is given in the Sect. 3.3.

The prediction and update operations are shown in Fig. 3.1, although the setup here is a little different from Chap. 2. We start with a finite sequence s_j of length 2^j. It is transformed into two sequences, each of length 2^{j-1}. They are denoted s_{j-1} and d_{j-1}, respectively. Let us explain the three steps in detail.

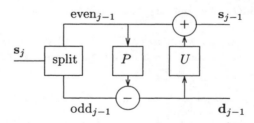

Fig. 3.1. The three steps in a lifting building block. Note that the minus means 'the signal from the left minus the signal from the top'

split The entries are sorted into the even and the odd entries. It is important to note that we do this only to explain the functionality of the algorithm. In (effective) implementations the entries are not moved or separated.

prediction If the signal contains some structure, then we can expect correlation between a sample and its nearest neighbors. In our first example the prediction is that the signal is constant. More elaborately, given the value at the sample number $2n$, we predict that the value at sample $2n+1$ is the same. We then replace the value at $2n+1$ with the correction to the prediction, which is the difference. In our notation this is (using the implementation given in (3.7))

$$d_{j-1}[n] = s_j[2n + 1] - s_j[2n] \ .$$

In general, the idea is to have a prediction procedure P and then compute

$$\mathbf{d}_{j-1} = \text{odd}_{j-1} - P(\text{even}_{j-1}) \ . \tag{3.9}$$

Thus in the $\mathbf{d}$ signal each entry is one odd sample minus some prediction based on a number of even samples.

update Given an even entry, we have predicted that the next odd entry has the same value, and stored the difference. We then update our even entry to reflect our knowledge of the signal. In the example above we replaced the even entry by the average. In our notation (and again using the implementation given in (3.7))

$$s_{j-1}[n] = s_j[2n] + d_{j-1}[n]/2 \ .$$

In general we decide on an updating procedure, and then compute

$$\mathbf{s}_{j-1} = \text{even}_{j-1} + U(\mathbf{d}_{j-1}) \ . \tag{3.10}$$

The algorithm described here is called one step lifting. It requires the choice of a prediction procedure P, and an update procedure U.

The discrete wavelet transform is obtained by combining a number of lifting steps. As in the example in Table 2.1 we keep the computed differences $\mathbf{d}_{j-1}$ and use the average sequence $\mathbf{s}_{j-1}$ as input for one more lifting step. This two step procedure is illustrated in Fig. 3.2.

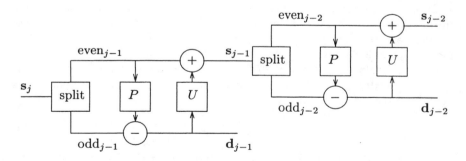

Fig. 3.2. Two step discrete wavelet transform

Starting with a signal $\mathbf{s}_j$ of length 2^j and repeating the transformations in the first example j times, we end up with a single number $s_0[0]$, which is easily seen to be the mean value of all entries in the original sequence. Taking $j = 3$ and using the same notation as in the tables in Chap. 2, then we see that the Table 2.1 is represented symbolically as Table 3.1. Now if we use the

Table 3.1. Notation for Table 2.1

$s_3[0]$	$s_3[1]$	$s_3[2]$	$s_3[3]$	$s_3[4]$	$s_3[5]$	$s_3[6]$	$s_3[7]$
$s_2[0]$	$s_2[1]$	$s_2[2]$	$s_2[3]$	$d_2[0]$	$d_2[1]$	$d_2[2]$	$d_2[3]$
$s_1[0]$	$s_1[1]$	$d_1[0]$	$d_1[1]$	$d_2[0]$	$d_2[1]$	$d_2[2]$	$d_2[3]$
$s_0[0]$	$d_0[0]$	$d_1[0]$	$d_1[1]$	$d_2[0]$	$d_2[1]$	$d_2[2]$	$d_2[3]$

'in place' procedure, and also record the intermediate steps, then we get the representation in Table 3.2. This table makes it evident that in implementing

Table 3.2. 'In place' representation for Table 3.1 with intermediate steps. Prediction steps are labeled with P, and update steps with U

$s_3[0]$	$s_3[1]$	$s_3[2]$	$s_3[3]$	$s_3[4]$	$s_3[5]$	$s_3[6]$	$s_3[7]$	
$s_3[0]$	$d_2[0]$	$s_3[2]$	$d_2[1]$	$s_3[4]$	$d_2[2]$	$s_3[6]$	$d_2[3]$	P
$s_2[0]$	$d_2[0]$	$s_2[1]$	$d_2[1]$	$s_2[2]$	$d_2[2]$	$s_2[3]$	$d_2[3]$	U
$s_2[0]$	$d_2[0]$	$d_1[0]$	$d_2[1]$	$s_2[2]$	$d_2[2]$	$d_1[1]$	$d_2[3]$	P
$s_1[0]$	$d_2[0]$	$d_1[0]$	$d_2[1]$	$s_1[1]$	$d_2[2]$	$d_1[1]$	$d_2[3]$	U
$s_1[0]$	$d_2[0]$	$d_1[0]$	$d_2[1]$	$d_0[0]$	$d_2[2]$	$d_1[1]$	$d_2[3]$	P
$s_0[0]$	$d_2[0]$	$d_1[0]$	$d_2[1]$	$d_0[0]$	$d_2[2]$	$d_1[1]$	$d_2[3]$	U

the procedure on the computer one has to be careful with the indices. For example, by inspecting the table carefully it is seen that one should step through the rows in steps of length 2, 4, and 8, while computing the s-values.

We have previously motivated the prediction operation with the reduction in dynamic range of the signal obtained in using differences rather than the original values, potentially leading to good compression of a signal. The update procedure has not yet been clearly motivated. The update performed in the first example in Chap. 2 was

$$s_{j-1}[n] = \tfrac{1}{2}\left(s_j[2n] + s_j[2n+1]\right). \tag{3.11}$$

It turns out that this operation preserves the mean value. The consequence is that all the **s** sequences have the same mean value. It is easy to verify in the case of the example in Table 2.1, since

$$\frac{56 + 40 + 8 + 24 + 48 + 48 + 40 + 16}{8}$$

$$= \frac{48 + 16 + 48 + 28}{4} = \frac{32 + 38}{2} = 35 \ .$$

It is not difficult to see that this hold for any s sequence of length 2^j. The mean value of such a sequence is

$$S = 2^{-j} \sum_{n=0}^{2^j - 1} s_j[n] \ .$$

Substituting (3.11) into this formula we get

$$\sum_{n=0}^{2^{j-1}-1} s_{j-1}[n] = \tfrac{1}{2} \sum_{n=0}^{2^{j-1}-1} (s_j[2n] + s_j[2n + 1]) = \tfrac{1}{2} \sum_{k=0}^{2^j-1} s_j[k] \ ,$$

which shows the result, since the signal s_j is twice as long as s_{j-1}. In particular, $s_0[0]$ equals the mean value of the original samples $s_j[0], \ldots, s_j[2^j - 1]$ (which in the first example was 35).

3.3 A Second Example

As mentioned earlier, there are many other possible prediction procedures, and update procedures. We give a second example. In our first example the prediction was correct for a constant signal. Now we want the prediction to be correct for a linear signal. We really mean an affine signal, but we stick to the commonly used term 'linear.' By a linear signal we mean a signal with the n-dependence of the form $s_j[n] = \alpha n + \beta$ (all the samples of the signal lie on a straight line). For a given odd entry $s_j[2n+1]$ we base the prediction on the two nearest even neighbors. The prediction is then $\frac{1}{2}(s_j[2n] + s_j[2n+2])$, since we want it to be correct for a linear signal. This value is the open circle in Fig. 3.3. The correction is the difference between what we predict the middle sample to be and what it actually is

$$d_{j-1}[n] = s_j[2n + 1] - \tfrac{1}{2}(s_j[2n] + s_j[2n + 2]) \ ,$$

and this difference is all we need to store. The principle is shown in Fig. 3.3.

We decide to base the procedure on the two most recently computed differences. We take it of the form

$$s_{j-1}[n] = s_j[2n] + A(d_{j-1}[n - 1] + d_{j-1}[n]) \ ,$$

where A is a constant to be determined. In the first example we had the property

$$\sum_n s_{j-1}[n] = \tfrac{1}{2} \sum_n s_j[n] \,. \tag{3.12}$$

We would like to have the same property here. Let us first rewrite the expression for $s_{j-1}[n]$ above,

$$
\begin{aligned}
s_{j-1}[n] &= s_j[2n] + Ad_{j-1}[n-1] + Ad_{j-1}[n] \\
&= s_j[2n] + A(s_j[2n-1] - \tfrac{1}{2}s_j[2n-2] - \tfrac{1}{2}s_j[2n]) \\
&\quad + A(s_j[2n+1] - \tfrac{1}{2}s_j[2n] - \tfrac{1}{2}s_j[2n+2]) \,.
\end{aligned}
$$

Using this expression, and gathering even and odd terms, we get

$$\sum_n s_{j-1}[n] = (1 - 2A) \sum_n s_j[2n] + 2A \sum_n s_j[2n+1] \,.$$

To satisfy (3.12) we must choose $A = \tfrac{1}{4}$. Summarizing, we have the following two steps

$$d_{j-1}[n] = s_j[2n+1] - \tfrac{1}{2}(s_j[2n] + s_j[2n+2]) \,, \tag{3.13}$$

$$s_{j-1}[n] = s_j[2n] + \tfrac{1}{4}(d_{j-1}[n-1] + d_{j-1}[n]) \,. \tag{3.14}$$

The transform in this example also has the property

$$\sum_n n s_{j-1}[n] = \tfrac{1}{2} \sum_n n s_j[n] \,. \tag{3.15}$$

We say that the transform preserves the *first moment* of the sequence. The average is also called the *zeroth moment* of the sequence.

In the above presentation we have simplified the notation by not specifying where the finite sequences start and end, thereby for the moment avoiding keeping track of the ranges of the variables. In other words, we have considered our finite sequences as infinite, adding zeroes before and after the given entries. In implementations one has to keep track of these things, but doing so now would obscure the simplicity of the lifting procedure. In later chapters we will deal with these problems in detail, see in particular Chap. 10.

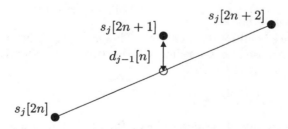

Fig. 3.3. The linear prediction

3.4 Lifting in General

We now look at the lifting procedure in general. Let us first look at how we can invert the lifting procedure. It is done by reversing the arrows and changing the signs. Thus the direct transform

$$\mathbf{d}_{j-1} = \mathrm{odd}_{j-1} - P(\mathrm{even}_{j-1})$$
$$\mathbf{s}_{j-1} = \mathrm{even}_{j-1} + U(\mathbf{d}_{j-1})$$

is inverted by the steps

$$\mathrm{even}_{j-1} = \mathbf{s}_{j-1} - U(\mathbf{d}_{j-1})$$
$$\mathrm{odd}_{j-1} = \mathbf{d}_{j-1} + P(\mathrm{even}_{j-1}) \ .$$

These steps are illustrated in Fig. 3.4. The last step, where the sequences even_{j-1} and odd_{j-1} are merged to form the sequence $\mathbf{s}_j$, is given to explain the algorithm. It is not performed in implementations, since the entries are not reordered. As an example, the inverse transform of (3.13) and (3.14) is

$$s_j[2n] = s_{j-1}[n] - \tfrac{1}{4}(d_{j-1}[n-1] + d_{j-1}[n]) \ , \qquad (3.16)$$
$$s_j[2n+1] = d_{j-1}[n] + \tfrac{1}{2}(s_j[2n] + s_j[2n+2]) \ . \qquad (3.17)$$

Looking at Fig. 3.4 once more, we see that the update step is reversed by the

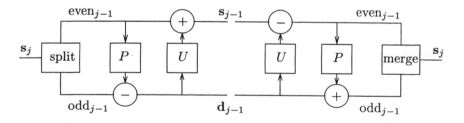

Fig. 3.4. Direct and inverse lifting step

same update step, but with subtraction instead of addition, and vice versa for the prediction step. Since each step is inverted separately, we can generalize in two ways. We can add further pairs of prediction and update steps, and we can add them singly. If we insist in having them in pairs (this is useful in the theory, see Chap. 12), we can always add an operation of either type which does nothing. As an illustration Fig. 3.5 shows a direct transform consisting of three pairs of prediction and update operations.

It turns out that this generalization is crucial in applications. There are many important transforms, where the steps do not occur in pairs. Here is an example, where there is a U operation followed by a P operation and

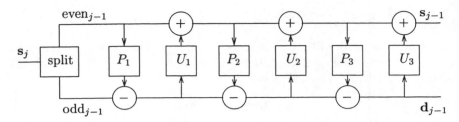

Fig. 3.5. Three lifting steps

another U operation. Furthermore, in the last two steps, in (3.21) and (3.22), we add a new type of operation which is called *normalization*, or sometimes rescaling. The resulting algorithm is applied to a signal $\{s_j[n]\}_{n \in \mathbf{Z}}$ as follows

$$s_{j-1}^{(1)}[n] = s_j[2n] + \sqrt{3}s_j[2n+1] \tag{3.18}$$

$$d_{j-1}^{(1)}[n] = s_j[2n+1] - \tfrac{1}{4}\sqrt{3}s_{j-1}^{(1)}[n] - \tfrac{1}{4}(\sqrt{3}-2)s_{j-1}^{(1)}[n-1] \tag{3.19}$$

$$s_{j-1}^{(2)}[n] = s_{j-1}^{(1)}[n] - d_{j-1}^{(1)}[n+1] \tag{3.20}$$

$$s_{j-1}[n] = \frac{\sqrt{3}-1}{\sqrt{2}}s_{j-1}^{(2)}[n] \tag{3.21}$$

$$d_{j-1}[n] = \frac{\sqrt{3}+1}{\sqrt{2}}d_{j-1}^{(1)}[n] \ . \tag{3.22}$$

Since there is more than one U operation, we have used superscripts on the **s** and **d** signals in order to tell them apart. Note that in the normalization steps we have

$$\frac{\sqrt{3}-1}{\sqrt{2}} \cdot \frac{\sqrt{3}+1}{\sqrt{2}} = 1 \ .$$

The reason for the normalization will become apparent in the next chapter, when we start doing computations. The algorithm above is one step in the discrete wavelet transform based on an important filter, which in the literature is often called Daubechies 4. The connection with filters will be explained in Chap. 7.

To find the inverse transform we have to use the prescription given above. We do the steps in reverse order and with the signs reversed. Thus the normalization is undone by multiplication by the inverse constants etc. The result is

$$d_{j-1}^{(1)}[n] = \frac{\sqrt{3}-1}{\sqrt{2}} d_{j-1}[n] \tag{3.23}$$

$$s_{j-1}^{(2)}[n] = \frac{\sqrt{3}+1}{\sqrt{2}} s_{j-1}[n] \tag{3.24}$$

$$s_{j-1}^{(1)}[n] = s_{j-1}^{(2)}[n] + d_{j-1}^{(1)}[n+1] \tag{3.25}$$

$$s_j[2n+1] = d_{j-1}^{(1)}[n] + \tfrac{1}{4}\sqrt{3}s_{j-1}^{(1)}[n] + \tfrac{1}{4}(\sqrt{3}-2)s_{j-1}^{(1)}[n-1] \tag{3.26}$$

$$s_j[2n] = s_{j-1}^{(1)}[n] - \sqrt{3}s_j[2n+1] . \tag{3.27}$$

This transform illustrates one of the problems that has to be faced in implementations. For example, to compute $d_{j-1}^{(1)}[0]$ we need to know $s_{j-1}^{(1)}[0]$ and $s_{j-1}^{(1)}[-1]$. But to compute $s_{j-1}^{(1)}[-1]$ one needs the values $s_j[-2]$ and $s_j[-1]$, which are not defined. The easiest solution to this problem is to use zero padding to get a sample at this index value (zero padding means that all undefined samples are defined to be 0). There exist other more sophisticated methods. This is the topic in Chap. 10.

Let us repeat our first example in the above notation. We also add a normalization step. In this form the transform is known as the Haar transform in the literature. The direct transform is

$$d_{j-1}^{(1)}[n] = s_j[2n+1] - s_j[2n] \tag{3.28}$$

$$s_{j-1}^{(1)}[n] = s_j[2n] + \tfrac{1}{2}d_{j-1}^{(1)}[n] \tag{3.29}$$

$$s_{j-1}[n] = \sqrt{2}s_{j-1}^{(1)}[n] \tag{3.30}$$

$$d_{j-1}[n] = \frac{1}{\sqrt{2}}d_{j-1}^{(1)}[n] \tag{3.31}$$

and the inverse transform is given by

$$d_{j-1}^{(1)}[n] = \sqrt{2}d_{j-1}[n] \tag{3.32}$$

$$s_{j-1}^{(1)}[n] = \frac{1}{\sqrt{2}}s_{j-1}[n] \tag{3.33}$$

$$s_j[2n] = s_{j-1}^{(1)}[n] - \tfrac{1}{2}d_{j-1}^{(1)}[n] \tag{3.34}$$

$$s_j[2n+1] = s_j[2n] + d_{j-1}^{(1)}[n] . \tag{3.35}$$

We note that this transform can be applied to a signal of length 2^j without using zero padding. It turns out to be the only transform with this property.

3.5 The Discrete Wavelet Transform in General

We now look at the discrete wavelet transform in the general framework established above. We postpone the boundary correction problem and assume that we have an infinite signal $\mathbf{s}_j = \{s_j[n]\}_{n\in\mathbf{Z}}$. The starting point is a transform (with a corresponding inverse transform) which takes as input a sequence $\mathbf{s}_j$ and produces as output two sequences $\mathbf{s}_{j-1}$ and $\mathbf{d}_{j-1}$ We will represent such a direct transform by the symbol $\mathbf{T}_a$ (subscript 'a' stands for analysis) and the inverse transform by the symbol $\mathbf{T}_s$ (subscript 's' stands for synthesis). In diagrams they will be represented as in Fig. 3.6. These are our fundamental building blocks.

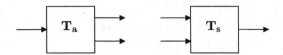

Fig. 3.6. Building blocks for DWT

The contents of the $\mathbf{T}_a$ box could be the direct Haar transform as given by (3.28)–(3.31), and the contents of the $\mathbf{T}_s$ box could be the inverse Haar transform as given by (3.32)–(3.35). Obviously, we must make sure to use the inverse transform corresponding to the applied direct transform. Otherwise, the results will be meaningless.

We can now combine these building blocks to get discrete wavelet transforms. We perform the transform over a certain number of scales j, meaning that we combine j of the building blocks as shown in Fig. 3.2 in the case of 2 scales, and in Fig. 3.7 in the case of 4 scales. In the latter figure we use the building block representation of the individual steps.

We use the symbol $W_a^{(j)}$ to denote a direct j scale discrete wavelet transform. The inverse is denoted by $W_s^{(j)}$. The result of the four scale transform is the transition

$$W_a^{(4)} : \mathbf{s}_j \quad \rightarrow \quad \mathbf{s}_{j-4}, \mathbf{d}_{j-4}, \mathbf{d}_{j-3}, \mathbf{d}_{j-2}, \mathbf{d}_{j-1} \ .$$

If we apply this four scale discrete wavelet transform to a signal of length 2^k, then the lengths on the right hand side are 2^{k-4}, 2^{k-4}, 2^{k-3}, 2^{k-2}, and 2^{k-1}, respectively. The sum of these five numbers is 2^k, as the reader easily verifies. The inverse four scale DWT is the transition

$$W_s^{(4)} : \mathbf{s}_{j-4}, \mathbf{d}_{j-4}, \mathbf{d}_{j-3}, \mathbf{d}_{j-2}, \mathbf{d}_{j-1} \quad \rightarrow \quad \mathbf{s}_j \ .$$

The diagram in Fig. 3.8 shows how it is computed. We use the term *scale* to describe how many times the building block $\mathbf{T}_a$ or $\mathbf{T}_s$ are applied in the decomposition of a signal. The word originates from the classical wavelet theory. The reader should note that we later, in Chap. 8, introduce the term *level*, in a context more general than the DWT. When this term is applied to a DWT decomposition, then the level is equal to the scale plus 1. The reader should not mix up the two terms.

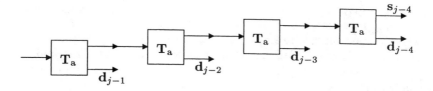

Fig. 3.7. DWT over four scales

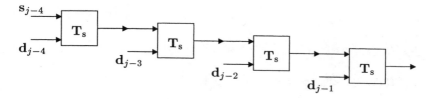

Fig. 3.8. Inverse DWT over four scales

3.6 Further Examples

We give some further examples of building blocks that one can use for constructing wavelet transforms. The example in Sect. 3.3 is part of a large family of so-called biorthogonal wavelet transforms. The transform given in (3.13) and (3.14) is known in the literature as CDF(2,2), since the 'inventors' of this transform are A. Cohen, I. Daubechies, and J.-C. Feauveau [2]. We give a larger part of the family below. The first step is in all three cases the same. The final normalization is also the same.

$$d_{j-1}^{(1)}[n] = s_j[2n+1] - \frac{1}{2}(s_j[2n] + s_j[2n+2]) \tag{3.36}$$

CDF(2,2) $\quad s_{j-1}^{(1)}[n] = s_j[2n] + \frac{1}{4}(d_{j-1}[n-1] + d_{j-1}[n]) \tag{3.37}$

CDF(2,4) $\quad s_{j-1}^{(1)}[n] = s_j[2n] - \frac{1}{64}(3d_{j-1}[n-2] - 19d_{j-1}[n-1]$
$$- 19d_{j-1}[n] + 3d_{j-1}[n+1]) \tag{3.38}$$

CDF(2,6) $\quad s_{j-1}^{(1)}[n] = s_j[2n] - \frac{1}{512}(-5d_{j-1}[n-3] + 39d_{j-1}[n-2]$
$$- 162d_{j-1}[n-1] - 162d_{j-1}[n]$$
$$+ 39d_{j-1}[n+1] - 5d_{j-1}[n+2]) \tag{3.39}$$

$$d_{j-1}[n] = \frac{1}{\sqrt{2}}d_{j-1}^{(1)}[n] \,, \tag{3.40}$$

$$s_{j-1}[n] = \sqrt{2}s_{j-1}^{(1)}[n] \,. \tag{3.41}$$

We have not given the formulas for the inverse transforms. They are obtained as above by reversing the arrows and changing the signs.

We give one further example of a family of three transforms. We have taken as an example transforms that start with an update step and a prediction step, which are common to all three. Again, at the end there is a normalization step.

$$s_{j-1}^{(1)}[n] = s_j[2n] - \frac{1}{3}s_j[2n-1] \tag{3.42}$$

$$d_{j-1}^{(1)}[n] = s_j[2n+1] - \frac{1}{8}(9s_{j-1}^{(1)}[n] + 3s_{j-1}^{(1)}[n+1]) \tag{3.43}$$

CDF(3,1) $$s_{j-1}^{(2)}[n] = s_{j-1}^{(1)}[n] + \frac{4}{9}d_{j-1}^{(1)}[n] \tag{3.44}$$

CDF(3,3) $$s_{j-1}^{(2)}[n] = s_{j-1}^{(1)}[n] + \frac{1}{36}(3d_{j-1}^{(1)}[n-1]$$
$$+ 16d_{j-1}^{(1)}[n] - 3d_{j-1}^{(1)}[n+1]) \tag{3.45}$$

CDF(3,5) $$s_{j-1}^{(2)}[n] = s_{j-1}^{(1)}[n] - \frac{1}{288}(5d_{j-1}^{(1)}[n-2] - 34d_{j-1}^{(1)}[n-1]$$
$$- 128d_{j-1}^{(1)}[n] + 34d_{j-1}^{(1)}[n+1]$$
$$- 5d_{j-1}^{(1)}[n+2]) \tag{3.46}$$

$$d_{j-1}[n] = \frac{\sqrt{2}}{3}d_{j-1}^{(1)}[n] \tag{3.47}$$

$$s_{j-1}[n] = \frac{3}{\sqrt{2}}s_{j-1}^{(2)}[n] . \tag{3.48}$$

The above formulas for the CDF(2,x) and CDF(3,x) families have been taken from the technical report [27]. Further examples can be found there.

Exercises

3.1 Verify that the CDF(2,2) transform, defined in (3.13) and (3.14), preserves the first moment, i.e. verify that (3.15) holds.

4. Analysis of Synthetic Signals

The discrete wavelet transform has been introduced in the previous two chapters. The general lifting scheme, as well as some examples of transforms, were presented, and we have seen one application to a signal with just 8 samples. In this chapter we will apply the transform to a number of synthetic signals, in order to gain some experience with the properties of the discrete wavelet transform. We will process some signals by transformation, followed by some alteration, followed by inverse transformation, as we did in Chap. 2 to the signal with 8 samples. Here we use significantly longer signals. As an example, we will show how this approach can be used to remove some of the noise in a signal. We will also give an example showing how to separate slow and fast variations in a signal.

The computations in this chapter have been performed using MATLAB. We have used the toolbox *Uvi_Wave* to perform the computations. See Chap. 14 for further information on software, and Chap. 13 for an introduction to MATLAB and *Uvi_Wave*. At the end of the chapter we give some exercises, which one should try after having read Sect. 13.1.

4.1 The Haar Transform

Our first examples are based on the Haar transform. The one scale direct Haar transform is given by equations (3.28)–(3.31), and its inverse by equations (3.32)–(3.35). We start with a very simple signal, given as a continuous signal by the sine function. More precisely, we take the function $\sin(4\pi t)$, with $0 \leq t \leq 1$. We now sample this signal at 512 equidistant points in $0 \leq t \leq 1$. This gives us a discrete signal s_9. The index 9 comes from the exponent $512 = 2^9$, as we described in Chap. 3. This signal is plotted in Fig. 4.1. We label the entries on the horizontal axis by sample index. Note that due to the density of the sampling the graph looks like the graph of the continuous function.

We want to perform a wavelet transform of this discrete signal. We choose to do this over three scales. If we order the entries in the transformed signal as in Table 2.1, then we get the result shown in Fig. 4.2. The ordering of the entries is s_6, d_6, d_7, d_8. At each index point we have plotted a vertical line of length equal to the value of the coefficient. It is not immediately obvious how

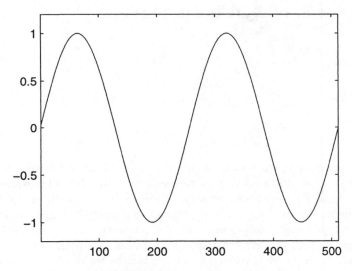

Fig. 4.1. The signal $\sin(4\pi t)$, 512 samples

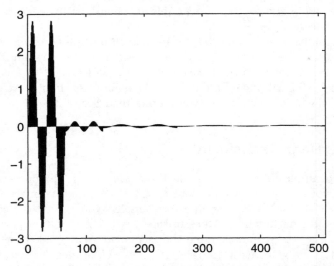

Fig. 4.2. The wavelet coefficients from the DWT of the signal in Fig. 4.1, using the Haar transform

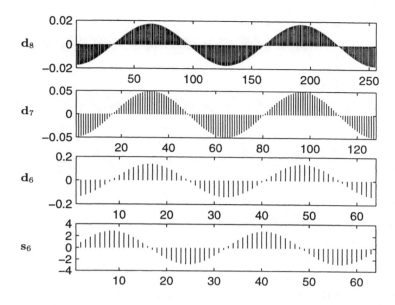

Fig. 4.3. The wavelet coefficients from Fig. 4.2 divided into scales, from the DWT of the signal in Fig. 4.1

one should interpret this graph. In Fig. 4.3 we have plotted the four parts separately. The top plot is of $\mathbf{d}_8$, followed by $\mathbf{d}_7$, $\mathbf{d}_6$, and $\mathbf{s}_6$. Note that each plot has its own axes, with different units. Again, these plots are not easy to interpret. We try a third approach.

We take the transformed signal, $\mathbf{s}_6, \mathbf{d}_6, \mathbf{d}_7, \mathbf{d}_8$, and then replace all entries except one with zeroes, i.e. sequences of the appropriate length consisting entirely of zeroes. For example, we can take $\mathbf{0}_6, \mathbf{d}_6, \mathbf{0}_7, \mathbf{0}_8$, where $\mathbf{0}_8$ is a signal of length $256 = 2^8$ with zero entries. We then invert this signal using the inverse three scale discrete wavelet transform based on (3.32)–(3.35). Schematically, it looks like

$$W_a^{(3)} : \mathbf{s}_9 \quad \rightarrow \quad \underbrace{\mathbf{s}_6, \mathbf{d}_6, \mathbf{d}_7, \mathbf{d}_8}$$
$$\downarrow$$
$$W_s^{(3)} : \overbrace{\mathbf{0}_6, \mathbf{d}_6, \mathbf{0}_7, \mathbf{0}_8} \quad \rightarrow \quad \mathbf{s}_9'$$

The result $\mathbf{s}_9'$ of this inversion is a signal with the property that if it were transformed with $W_a^{(3)}$, the result would be precisely the signal $\mathbf{0}_6, \mathbf{d}_6, \mathbf{0}_7, \mathbf{0}_8$. Hence $\mathbf{s}_9'$ contains all information on the coefficients on the third scale. The four possible plots are given in Fig. 4.4. The top plot is the inversion of $\mathbf{0}_6, \mathbf{0}_6, \mathbf{0}_7, \mathbf{d}_8$ followed by $\mathbf{0}_6, \mathbf{0}_6, \mathbf{d}_7, \mathbf{0}_8$, and $\mathbf{0}_6, \mathbf{d}_6, \mathbf{0}_7, \mathbf{0}_8$, and finally at the bottom $\mathbf{s}_6, \mathbf{0}_6, \mathbf{0}_7, \mathbf{0}_8$. This representation, where the contributions are separated as described, is called the *multiresolution* representation of the signal

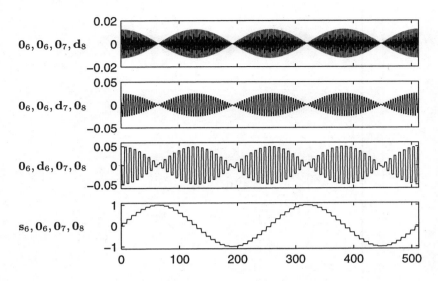

Fig. 4.4. DWT of the signal in Fig. 4.1, Haar transform, multiresolution representation, separate plots

(in this case over three scales). The plots in Fig. 4.4 correspond to our intuition associated with repeated means and differences. The bottom plot in Fig. 4.4 could also have been obtained by computing the mean of 8 successive samples, and then replacing each of these 8 samples by their mean value. Thus each mean value is repeated 8 times in the plot.

If we invert the transform, we will get back the original signal. In Fig. 4.5 we have plotted the inverted signal, and the difference between this signal and the original signal. We see that the differences are of magnitude 10^{-15}, corresponding to the precision of the MATLAB calculations.

We have now presented a way of visualizing the effect of a DWT over a finite number of scales. We will then perform some experiments with synthetic signals. As a first example we add an impulse to our sine signal. We change the value at sample number 200 to the value 2. We have plotted the three scale representation in Fig. 4.6. We see that the impulse can be localized in the component d_8, and in the averaged signal s_6 the impulse has almost disappeared. Here we have used the very simple Haar transform. By using other transforms one can get better results.

Let us now show how we can reduce noise in a signal by processing it in the DWT representation. We take again the sine signal plus an impulse, and add some noise. The signal is given in Fig. 4.7. The multiresolution representation is given in Fig. 4.8 The objective now is to remove the noise from the signals. We will try to do this by processing the signal as follows. In the transformed representation, s_6, d_6, d_7, d_8, we leave unchanged the largest

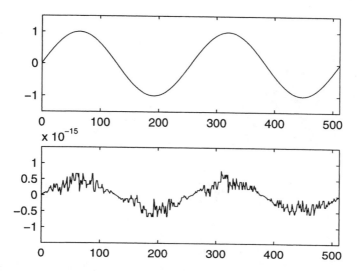

Fig. 4.5. Top: Inverse DWT of the signal in Fig. 4.1. Bottom: Difference between inverted and original signal

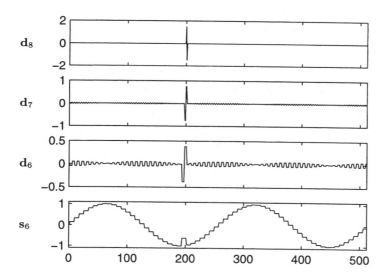

Fig. 4.6. Multiresolution representation of sine plus impulse at 200, Haar transform

10% of the coefficients, and change the remaining 90% to zero. We then apply the inverse transform to this altered signal. The result is shown in Fig. 4.9. We see that it is possible to recognize both the impulse and the sine signal, but the sine signal has undergone considerable changes. The next section

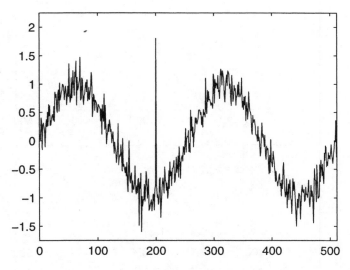

Fig. 4.7. Sine plus impulse at 200 plus noise

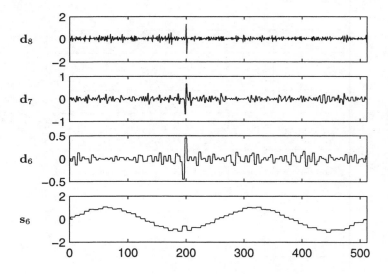

Fig. 4.8. Sine plus impulse at 200 plus noise, Haar transform, multiresolution representation

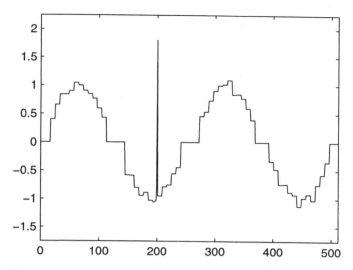

Fig. 4.9. Sine plus impulse at 200 plus noise, reconstruction based on the largest 10% of the coefficients, Haar transform

shows that these results can be improved by choosing a more complicated transform.

4.2 The CDF(2,2) Transform

We will now perform experiments with the DWT based on the building block CDF(2,2), as it was defined in Sect. 3.3. We will continue with the noise reduction example from the previous section. In Fig. 4.10 we have given the multiresolution representation, using the new building block for the DWT. In Fig. 4.11 we have shown reconstruction based on the 15% and the 10% largest coefficients in the transformed signal. The result is much better than the one obtained using the Haar transform. Let us note that there exists an extensive theory on noise removal, including very sophisticated applications of the DWT, but it is beyond the scope of this book.

As a second example we show how to separate fast and slow variations in a signal. We take the function

$$\log(2 + \sin(3\pi\sqrt{t})), \qquad 0 \le t \le 1, \tag{4.1}$$

and sample its values in 1024 points, at $1/1024$, $2/1024$, ... , $1024/1024$. Then we change the values at $1/1024$, $33/1024$, $65/1024$, etc. by adding 2 to the computed values. This signal has been plotted in Fig. 4.12. We will now try to separate the slow variation and the sharp peaks in the signal. We take

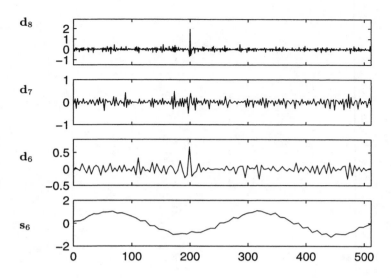

Fig. 4.10. Sine plus impulse at 200 plus noise, multiresolution representation, CDF(2,2), three scales

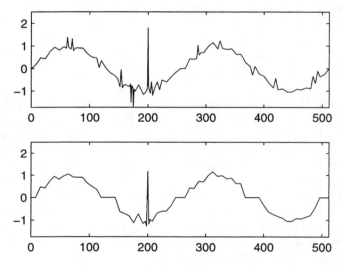

Fig. 4.11. Sine plus impulse at 200 plus noise, CDF(2,2), three scales, top reconstruction based on 15% largest coefficients, bottom based on 10% largest coefficients

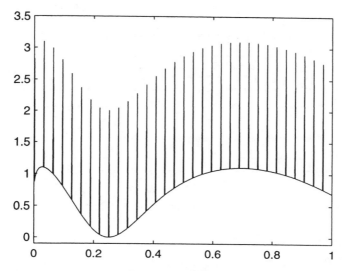

Fig. 4.12. Plot of the function $\log(2+\sin(3\pi\sqrt{t}))$ plus 2 at points 1/1024, 33/1024, 65/1024, etc.

a multiresolution representation over six scales, as shown in Fig. 4.13. We see from the bottom graph in the figure that we have succeeded in removing the sharp peaks in that part of the representation. In Fig. 4.14 we have plotted this part separately. Except close to the end points of the interval, this is the slow variation. We subtract this part from the original signal and obtain Fig. 4.15. In these two figures we have used the variable t on the horizontal axis. Figure 4.15 shows that except for problems at the edges we have succeeded in isolating the rapid variations, without broadening the sharp peaks in the rapidly varying part. This example is not only of theoretical interest, but can also be applied to for example ECG signals.

Exercises

All exercises below require access to MATLAB and *Uvi_Wave* (or some other wavelet toolbox), and some knowledge of their use. You should read Sect. 13.1 before trying to solve these exercises.

4.1 Go through the examples in this chapter, using MATLAB and *Uvi_Wave*.

4.2 Carry out experiments on the computer with noise reduction. Vary the number of coefficients retained, and plot the different reconstructions. Discuss the results.

4.3 Find the multiresolution representation of a chirp (i.e. a signal obtained from sampling $\sin(t^2)$).

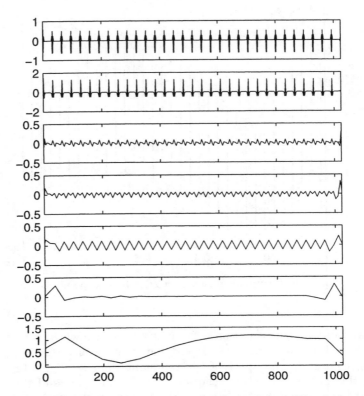

Fig. 4.13. Multiresolution representation of the signal from Fig. 4.12, six scales, CDF(2,2)

4.4 Find the multiresolution representation of a signal obtained by sampling the function

$$f(t) = \begin{cases} \sin(4\pi t) & \text{for } 0 \le t < \frac{1}{4}, \\ 1 + \sin(4\pi t) & \text{for } \frac{1}{4} \le t < \frac{3}{4}, \\ \sin(4\pi t) & \text{for } \frac{3}{4} \le t \le 1. \end{cases}$$

Add noise, and try out noise removal, using both the Haar transform, and the CDF(2,2) transform.

4.5 (For readers with sufficient background in signal analysis.) Try to separate the low and high frequencies in the signal in Fig. 4.12 by a low pass filtering (use for example a low order Butterworth filter). Compare the result to Fig. 4.14. Subtract the low pass filtered signal from the original and compare the result to Fig. 4.15.

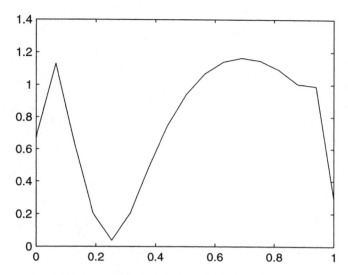

Fig. 4.14. Bottom graph in Fig. 4.13, fitted to $0 \leq t \leq 1$

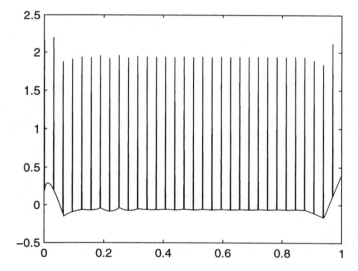

Fig. 4.15. Signal from Fig. 4.12, with slow variations removed

5. Interpretation

In this chapter we start with an interpretation of the discrete wavelet transform based on the Haar building block. Then we will give interpretations of wavelet transforms based on more general building blocks. The last part of this chapter can be omitted on a first reading.

Our presentation in this chapter is incomplete. We state some results from the general theory, and illustrate them with explicit computations. But we do not discuss the general theory in detail, since this requires a mathematical background that we do not assume our readers possess.

Understanding the results in this chapter will be much easier, if one carries out extensive computer experiments, as we suggest in the exercises at the end of the chapter. The necessary background and explanations can be found in Sect. 13.1. It can be read now, since it does not depend on the following chapters.

5.1 The First Example

We go back to the first example, discussed in Chap. 2. Again we begin with a signal of length $8 = 2^3$. This means that we can work at up to three scales. For the moment we order the entries in the transforms as in Table 2.1. The goal is to give an interpretation of the transformed signal, i.e. the last row in Table 2.1. What does this transformed signal reveal about the given signal? How can we interpret these numbers? One way to answer these questions is to start with the bottom row, consisting of zeroes except at one entry, and then inversely transform this signal. In other words, we find the signal whose transform consists of zeroes except for a 1 at one entry. We keep the ordering s_0, d_0, d_1, d_2. The results of the first three computations are given in Tables 5.1–5.3. The remaining cases are left as exercises for the reader. Note that we continue with the conventions from Chap. 2, in other words, we omit the normalization steps given in equations (3.30) and (3.31). This does not change the interpretation, and the tables below become easier to understand. Later we will explain why normalization is needed.

The results of these eight computations can be represented using notation from linear algebra. A signal of length 8 is a vector in the vector space $\mathbf{R}^8$. The process described above of reconstructing the signal, whose transform is

Table 5.1. Reconstruction based on $[1, 0, 0, 0, 0, 0, 0, 0]$

1	1	1	1	1	1	1	1
1	1	1	1	0	0	0	0
1	1	0	0	0	0	0	0
1	0	0	0	0	0	0	0

Table 5.2. Reconstruction based on $[0, 1, 0, 0, 0, 0, 0, 0]$

1	1	1	1	−1	−1	−1	−1
1	1	−1	−1	0	0	0	0
1	−1	0	0	0	0	0	0
0	1	0	0	0	0	0	0

Table 5.3. Reconstruction based on $[0, 0, 1, 0, 0, 0, 0, 0]$

1	1	−1	−1	0	0	0	0
1	−1	0	0	0	0	0	0
0	0	1	0	0	0	0	0
0	0	1	0	0	0	0	0

one of the canonical basis vectors in $\mathbf{R}^8$, is the same as finding the columns in the matrix (with respect to the canonical basis) of the three scale synthesis transform $W_s^{(3)}$. This matrix, which we denote by $\mathbf{W}_s^{(3)}$, is shown in (5.1).

$$
\mathbf{W}_s^{(3)} = \begin{bmatrix}
1 & 1 & 1 & 0 & 1 & 0 & 0 & 0 \\
1 & 1 & 1 & 0 & -1 & 0 & 0 & 0 \\
1 & 1 & -1 & 0 & 0 & 1 & 0 & 0 \\
1 & 1 & -1 & 0 & 0 & -1 & 0 & 0 \\
1 & -1 & 0 & 1 & 0 & 0 & 1 & 0 \\
1 & -1 & 0 & 1 & 0 & 0 & -1 & 0 \\
1 & -1 & 0 & -1 & 0 & 0 & 0 & 1 \\
1 & -1 & 0 & -1 & 0 & 0 & 0 & -1
\end{bmatrix}.
\tag{5.1}
$$

The first row in Table 5.1 is the transpose of the first column in (5.1), and so on. Applying this matrix to any length 8 signal performs the inverse Haar transform. For example, multiplying it with the fourth row of Table 2.3 on p. 9 (regarded as a column vector) produces the first row of that same table.

The matrix of the direct three scale transform is obtained by computing the transforms of the eight canonical basis vectors in $\mathbf{R}^8$. In other words, we start with the signal $[1, 0, 0, 0, 0, 0, 0, 0]$ and carry out the transform as shown in Table 5.4, and analogously for the remaining seven basis vectors. The result is the matrix of the direct, or analysis, transform.

Table 5.4. Direct transform of first basis vector

1	0	0	0	0	0	0	0
$\frac{1}{2}$	0	0	0	$\frac{1}{2}$	0	0	0
$\frac{1}{4}$	0	$\frac{1}{4}$	0	$\frac{1}{2}$	0	0	0
$\frac{1}{8}$	$\frac{1}{8}$	$\frac{1}{4}$	0	$\frac{1}{2}$	0	0	0

$$
\mathbf{W}_a^{(3)} = \begin{bmatrix}
\frac{1}{8} & \frac{1}{8} & \frac{1}{8} & \frac{1}{8} & \frac{1}{8} & \frac{1}{8} & \frac{1}{8} & \frac{1}{8} \\
\frac{1}{8} & \frac{1}{8} & \frac{1}{8} & \frac{1}{8} & -\frac{1}{8} & -\frac{1}{8} & -\frac{1}{8} & -\frac{1}{8} \\
\frac{1}{4} & \frac{1}{4} & -\frac{1}{4} & -\frac{1}{4} & 0 & 0 & 0 & 0 \\
0 & 0 & 0 & 0 & \frac{1}{4} & \frac{1}{4} & -\frac{1}{4} & -\frac{1}{4} \\
\frac{1}{2} & -\frac{1}{2} & 0 & 0 & 0 & 0 & 0 & 0 \\
0 & 0 & \frac{1}{2} & -\frac{1}{2} & 0 & 0 & 0 & 0 \\
0 & 0 & 0 & 0 & \frac{1}{2} & -\frac{1}{2} & 0 & 0 \\
0 & 0 & 0 & 0 & 0 & 0 & \frac{1}{2} & -\frac{1}{2}
\end{bmatrix}.
\tag{5.2}
$$

Multiplying the matrices we find $\mathbf{W}_s^{(3)} \cdot \mathbf{W}_a^{(3)} = \mathbf{I}$ and $\mathbf{W}_a^{(3)} \cdot \mathbf{W}_s^{(3)} = \mathbf{I}$, where $\mathbf{I}$ denotes the 8×8 unit matrix. This is the linear algebra formulation of perfect reconstruction, or of the invertibility of the three scale transform.

It is clear that analogous constructions can be carried out for signals of length 2^j and transforms to all scales k, $k = 1, \ldots, j$. The linear algebra point of view is useful in understanding the theory, but if one is trying to carry out numerical computations, then it is a bad idea to use the matrix formulation. The direct k scale wavelet transform using the lifting steps requires a number of operations (additions, multiplications, etc.) on the computer, which is proportional to the length L of the signal. If we perform the transform using its matrix, then in general a number of operations proportional to L^2 is needed.

We can learn one more thing from this example. Let us look at the direct transforms

$$T_a : \quad [1,0,0,0,0,0,0,0] \to [\tfrac{1}{8}, \tfrac{1}{8}, \tfrac{1}{4}, 0, \tfrac{1}{2}, 0, 0, 0] \,,$$

and

$$T_a : \quad [0,0,0,0,1,0,0,0] \to [\tfrac{1}{8}, -\tfrac{1}{8}, 0, \tfrac{1}{4}, 0, 0, \tfrac{1}{2}, 0] \,.$$

The second signal is the first signal translated four units in time, but the transforms look rather different. Actually, this is one of the reasons why wavelet transforms can be used to localize events in time, as illustrated in some of the simple examples in Chap. 4. Readers familiar with Fourier analysis will see that the wavelet transform is quite different from the Fourier transform with respect to translations in time.

5.2 Further Results on the Haar Transform

In the next two subsections we describe some further results on the Haar transform. They are of a more advanced nature than the other results in this chapter, but are needed for a deeper understanding of the DWT. The remaining parts of this chapter can be omitted on a first reading.

5.2.1 Normalization of the Transform

Let us now explain why we normalize the Haar transform in the steps (3.30)–(3.31). This explanation requires a bit more mathematics than the rest of the chapter. We use the space $\ell^2(\mathbf{Z})$ of signal of finite energy, introduced in Chap. 3. The energy in a signal $\mathbf{s}$ is measured by the quantity

$$\|\mathbf{s}\|^2 = \sum_{n \in \mathbf{Z}} |s[n]|^2 \ . \tag{5.3}$$

The square root of the energy, denoted by $\|\mathbf{s}\|$, is called the *norm* of the vector $\mathbf{s} \in \ell^2(\mathbf{Z})$. We now compute the norms of the signals in the first and last row of Table 5.1

$$\|[1,1,1,1,1,1,1,1]\| = \sqrt{8} \ ,$$
$$\|[1,0,0,0,0,0,0,0]\| = 1 \ .$$

Recall that we identify finite signals with infinite signals by adding zeroes. If we carry out the same computation with a signal of length 2^N, consisting of a 1 followed by $2^N - 1$ zeroes, then we find that the inverse N scale Haar transform (computed as in Table 5.1) of this signal is a vector of length 2^N, all of whose entries are ones. This vector has norm $2^{N/2}$. This means that the norm grows exponentially with N. Such growth can easily lead to numerical instability. It is to avoid such instability that one chooses to normalize the Haar building blocks. The consequence of the normalization is that we have

$$\|W_{\mathrm{a}}^{(k),\mathrm{haar,norm}}\mathbf{x}\| = \|\mathbf{x}\| \quad \text{and} \quad \|W_{\mathrm{s}}^{(k),\mathrm{haar,norm}}\mathbf{x}\| = \|\mathbf{x}\|$$

at any scale k, compatible with the length of the signal. This result applies to both finite length signals and infinite length signals of finite energy, i.e. for all $\mathbf{x} \in \ell^2(\mathbf{Z})$.

It is not always possible to obtain this nice property, but we will at least require that – after normalization – the norm of the signal, and the norm of its direct and inverse transforms, have the same order of magnitude. This is expressed by requiring the existence of constants $A, B, \tilde{A}, \tilde{B}$, such that

$$A\|\mathbf{x}\| \leq \|T_{\mathrm{a}}\mathbf{x}\| \leq B\|\mathbf{x}\| \ , \tag{5.4}$$

$$\tilde{A}\|\mathbf{x}\| \leq \|T_{\mathrm{s}}\mathbf{x}\| \leq \tilde{B}\|\mathbf{x}\| \ , \tag{5.5}$$

for all $\mathbf{x} \in \ell^2(\mathbf{Z})$. All the building blocks given in Chap. 3 have this property. In particular, the Haar and Daubechies 4 transform have $A = B = \tilde{A} = \tilde{B} = 1$. Note that we here use the generality of our transform. The transforms T_a and T_s can be applied to *any* vector $\mathbf{x} \in \ell^2(\mathbf{Z})$.

Since the transforms $W_a^{(N)}$ and $W_s^{(N)}$ are obtained by iterating the building blocks, similar estimates hold for these transforms, with constants that may depend on the scale N.

5.2.2 A Special Property of the Haar Transform

In this section we describe a special property of the Haar transform. We start by looking at the inversion results summarized in Tables 5.1–5.3, and in the matrix (5.1). We can obtain analogous results using signals of length 2^N and performing inverse N scale Haar transform (not normalized) on the 2^N signals $[1, 0, \ldots, 0]$, $[0, 1, \ldots, 0]$, $\ldots$, $[0, 0, \ldots, 1]$. For example, we find that the inverse transform of

$$\underbrace{[1, 0, \ldots, 0]}_{2^N \text{ entries}} \quad \text{is} \quad \underbrace{[1, 1, \ldots, 1]}_{2^N \text{ entries}}, \tag{5.6}$$

and that the inverse transform of

$$\underbrace{[0, 1, \ldots, 0]}_{2^N \text{ entries}} \quad \text{is} \quad [\underbrace{1, 1, \ldots, 1}_{2^{N-1} \text{ entries}}, \underbrace{-1, -1, \ldots, -1}_{2^{N-1} \text{ entries}}]. \tag{5.7}$$

Finally, we compute as in Table 5.3 that the inverse transform of

$$\underbrace{[0, 0, 1, \ldots, 0]}_{2^N \text{ entries}} \quad \text{is} \quad [\underbrace{1, 1, \ldots, 1}_{2^{N-2} \text{ entries}}, \underbrace{-1, -1, \ldots, -1}_{2^{N-2} \text{ entries}}, \underbrace{0, 0, \ldots, 0}_{2^{N-1} \text{ entries}}]. \tag{5.8}$$

There is a pattern in these computations which can be understood as follows. We can imagine that all these signals come from continuous signals, which have been sampled. We choose the time interval to be the interval $[0, 1]$. Then the sampling points are $1 \cdot 2^{-N}$, $2 \cdot 2^{-N}$, $3 \cdot 2^{-N}$, $\ldots$, $2^N \cdot 2^{-N}$. For the first signal (5.6) we see that it is obtained by sampling the function

$$h_0(t) = 1, \quad t \in [0, 1],$$

which is *independent* of N. The second signal (5.7) is obtained from the function

$$h_1(t) = \begin{cases} 1, & t \in [0, \frac{1}{2}], \\ -1, & t \in]\frac{1}{2}, 1], \end{cases}$$

again with a function independent of N. The third signal (5.8) is obtained from sampling the function

$$h_2(t) = \begin{cases} 1, & t \in [0, \frac{1}{4}], \\ -1, & t \in]\frac{1}{4}, \frac{1}{2}], \\ 0, & t \in]\frac{1}{2}, 1]. \end{cases}$$

The pattern is described as follows. We define the function

$$h(t) = \begin{cases} 1, & t \in [0, \frac{1}{2}], \\ -1, & t \in]\frac{1}{2}, 1]. \end{cases} \tag{5.9}$$

Consider now the following way of writing the positive integers. For $n = 1, 2, 3, \ldots$ we write $n = k + 2^j$ with $j \geq 0$ and $0 \leq k < 2^j$. Each integer has a unique decomposition. Sample computations are shown in Table 5.5.

Table 5.5. Index computation for Haar basis functions

j	0	1	1	2	2	2	2	3	3	3	3	3	$\cdots$
k	0	0	1	0	1	2	3	0	1	2	3	4	$\cdots$
$n = k + 2^j$	1	2	3	4	5	6	7	8	9	10	11	12	$\cdots$

The general function (5.9) is then described by

$$h_n(t) = h(2^j t - k), \quad t \in [0, 1], \quad n = 1, 2, 3, \ldots,$$

with n, j, k related as just described. For a given N, the 2^N vectors described above are obtained by sampling $h_0, h_1, \ldots, h_{2^N - 1}$. The important thing to notice is that all functions are obtained from just two different functions. The function h_0, which will be called the *scaling function*, and the function h, which will be called the *wavelet*. The functions h_n, $n = 1, 2, 3, \ldots$, are obtained from the single function h by *scaling*, determined by j, and *translation*, determined by k.

The functions defined above are called the Haar (basis) functions. In Fig. 5.1 we have plotted the first eight functions. These eight functions are the ones that after sampling lead to the columns in the matrix (5.1).

We should emphasize that the above computations are performed with the non-normalized transform. If we introduce normalization, then we have to use the functions

$$h_n^{\text{norm}}(t) = 2^{j/2} h(2^j t - k), \quad t \in [0, 1], \quad n = 1, 2, 3, \ldots.$$

Let us also look at the role of the scaling function h_0 above. In the example we transformed a signal of length 8 three scales down. We could have chosen only to go two scales down. In this case the signal s_3 is transformed into s_1, d_1, and d_2, of length 2, 2, and 4, respectively. We can perform the inversion of the eight possible unit vectors, as above. For the 1 entries in d_1 and d_2 the

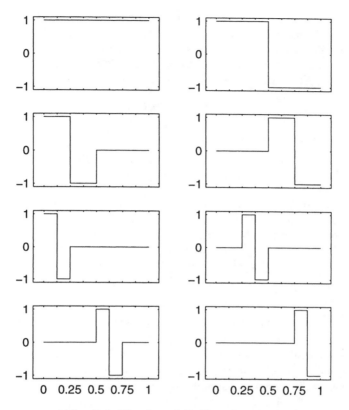

Fig. 5.1. The first eight Haar functions

Table 5.6. Two scale reconstruction for first two unit vectors

1	1	1	1	0	0	0	0
1	1	0	0	0	0	0	0
1	0	0	0	0	0	0	0

0	0	0	0	1	1	1	1
0	0	1	1	0	0	0	0
0	1	0	0	0	0	0	0

results are unchanged, as one can see from the first three lines in Table 5.3. The results for the two cases with ones in s_1 are given in Table 5.6 These two vectors can be obtained by sampling the two functions $h_0(2t)$ and $h_0(2t-1)$.

In general, one finds for a given signal of length 2^N, transformed down a number of scales k, $1 \le k \le N$, results analogous to those above. Thus one gets a basis of $\mathbf{R}^{2^N}$ consisting of vectors obtained by sampling scaled and translated functions $h_0(2^n t - m)$ and $h(2^n t - m)$. Here n and m run through values determined by the scale considered. As mentioned above, h_0 is called the scaling function and h the wavelet.

Let us now try to give an interpretation of the direct transform in terms of these functions. Let us again take $N = 3$, i.e. signals of length 8. The eight functions $h_0, \ldots, h_7$, sampled at $\frac{1}{8}, \frac{2}{8}, \ldots, \frac{8}{8}$, give the vectors, whose three scale direct Haar transform are the eight basis vectors $[1, 0, \ldots, 0], \ldots,$ $[0, 0, \ldots, 1]$. We will here use the notation

$$\mathbf{e}_0 = [1, 0, 0, 0, 0, 0, 0, 0],$$
$$\mathbf{e}_1 = [0, 1, 0, 0, 0, 0, 0, 0],$$
$$\vdots$$
$$\mathbf{e}_7 = [0, 0, 0, 0, 0, 0, 0, 1].$$

The eight sampled functions will be denoted by $\mathbf{h}_0, \ldots, \mathbf{h}_7$. They are given by the columns in the matrix (5.1), as before. The transform relationships can be expressed by the equations

$$\mathbf{W}_a^{(3)}(\mathbf{h}_n) = \mathbf{e}_n, \quad n = 0, 1, \ldots, 7,$$

and

$$\mathbf{W}_s^{(3)}(\mathbf{e}_n) = \mathbf{h}_n, \quad n = 0, 1, \ldots, 7.$$

Note that we here have to take the transpose of the vectors $\mathbf{e}_k$ defined above, apply the matrix, and then transpose the result to get the row vectors $\mathbf{h}_k$.

Now let us take a general signal $\mathbf{x}$ of length 8, and let $\mathbf{y} = \mathbf{W}_a^{(3)}(\mathbf{x})$ denote the direct transform. Since both the direct and inverse transforms are linear (preserve linear combinations), and since

$$\mathbf{y} = \sum_{n=0}^{7} y[n]\mathbf{e}_n,$$

we have the relation

$$\mathbf{x} = \mathbf{W}_s^{(3)}(\mathbf{y}) = \sum_{n=0}^{7} y[n]\mathbf{h}_n. \tag{5.10}$$

Thus the direct transform $\mathbf{W}_a^{(3)}$ applied to $\mathbf{x}$ yields a set of coefficients $\mathbf{y}$, with the property that the original signal $\mathbf{x}$ is represented as a superposition of the elementary signals $\mathbf{h}_0, \ldots, \mathbf{h}_7$, as shown in (5.10). The weight of each elementary signal is given by the corresponding transform coefficient $y[n]$.

In the next section we will see that *approximately* the same pattern can be found in more general transforms, although it is not so easy to obtain as in the case of the Haar transform.

5.3 Interpretation of General Discrete Wavelet Transforms

In this section we give some further examples of the procedure used in the previous sections, and then state the general result. This section is rather incomplete, since a complete treatment of these results requires a considerable mathematical background.

5.3.1 Some Examples

We will start by repeating the computations in Sect. 5.1 using – instead of the inverse Haar transform – the transform given by (3.23)–(3.27), which we call the inverse Daubechies 4 transform. We take as an example the vector $[0, 0, 0, 0, 0, 1, 0, 0]$ of length 8 and perform a three scale inverse transform. The entries are plotted against the points $\frac{1}{8}, \frac{2}{8}, \ldots, \frac{8}{8}$ on the t-interval $[0, 1]$ in Fig. 5.2. This figure contains very little information. But let us now repeat

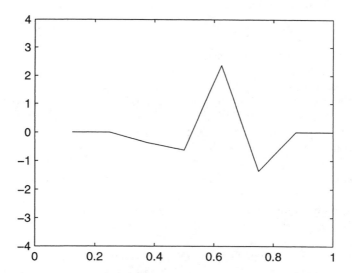

Fig. 5.2. Inverse Daubechies 4 of $[0, 0, 0, 0, 0, 1, 0, 0]$ over three scales, rescaled

the procedure for vectors of length 8, 32, 128, and 512, applied to a vector with a single 1 as its sixth entry. We fit each transform to the interval $[0, 1]$. This requires that we rescale the values of the transform by $2^{k/2}$, $k = 3, 5, 7, 9$. The result is shown in Fig. 5.3. We recall that the inverse Daubechies 4 transform includes the normalization step. This figure shows that the graphs rapidly approach a limiting graph, as we increase the length of the vector. This is a

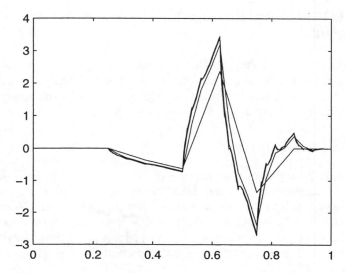

Fig. 5.3. Inverse Daubechies 4 of sixth basis vector, length 8, 32, 128 and 512, rescaled

result that can be established rigorously, but it is not easy to do so, and it is way beyond the scope of this book.

One can interpret the limiting function in Fig. 5.3 as a function whose values, sampled at appropriate points, represent the entries in the inverse transform of a vector of length 2^N, with a single 1 as its sixth entry. For N just moderately large, say $N = 12$, this is a very good approximation to the actual value. See Fig. 5.4 for the result for $N = 12$, i.e. a vector of length 4096.

For all other basis vectors, except $[1, 0, 0, \dots, 0]$, one gets similar results in the sense that the graph has the same form, but will be scaled and/or translated. The underlying function is also here called the wavelet. For the vector $[1, 0, 0, \dots, 0]$ one gets a different graph. The underlying function is again called the scaling function.

The theory shows that if one chooses to transform a signal of length 2^N to a scale k, then the inverse transform of the unit vectors with ones at places 1 to 2^{N-k} will be approximations to translated copies of the scaling function.

Let us repeat these computations for the inverse transform CDF(2,2) from Sect. 3.3 (the formulas for the inverse are given in Sect. 3.4), and at the same time illustrate how the wavelet is translated depending on the placement of the 1 in the otherwise 0 vector. An example is given in Fig. 5.5. The difference in the graphs in Fig. 5.5 and Fig. 5.4 is striking. It reflects the result that the Daubechies 4 wavelet has very little regularity (it is not differentiable), whereas the other wavelet is a piecewise linear function.

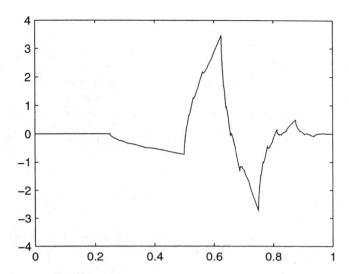

Fig. 5.4. Inverse Daubechies 4 of sixth basis vector, length 4096, rescaled. The result is the Daubechies 4 wavelet

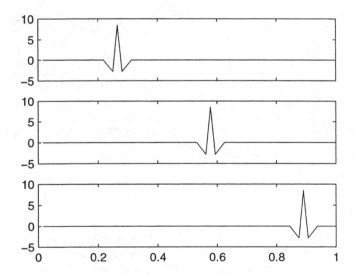

Fig. 5.5. Inverse CDF(2,2) of three basis vectors of length 64, entry 40, or 50, or 60, equal to 1 and the remaining entries equal to zero. The result is the same function (the wavelet) with different translations

Finally, if we try to find the graphs of the scaling function and wavelet underlying the direct CDF(2,2) transform, then we get the graphs in Fig. 5.6 and Fig. 5.7. These functions are quite complicated. These figures have been generated by taking $N = 16$ and a transform of $k = 12$ scales. To generate the scaling function we have taken the transform of a vector with a one at place 8. For the wavelet we have taken the one at place 24. It is interesting to see that while the analysis wavelet and scaling function are very simple functions (we have not shown the scaling function of CDF(2,2), see Exer. 5.6), the inverse of that same transform (synthesis) have some rather complex wavelet and scaling functions.

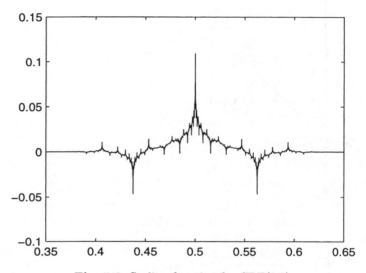

Fig. 5.6. Scaling function for CDF(2,2)

5.3.2 The General Case

The above computations may lead us to the conclusion that there are just two functions underlying the direct transform, and another two functions underlying the inverse transform, in the sense that if we take sufficiently long vectors, say 2^N, and perform a k scale transform, with k large, then we get values that are sampled values of one of the underlying functions. More precisely, inverse transforms of unit vectors with a one in places from 1 to 2^{N-k} yield translated copies of the scaling function. Inverse transforms of unit vectors with a one in places from $2^{N-k} + 1$ to 2^{N-k+1} yield translated copies of the wavelet. Finally, inverse transforms of unit vectors with a one at places from $2^{N-k+1} + 1$ to 2^N yield scaled and translated copies of the wavelet.

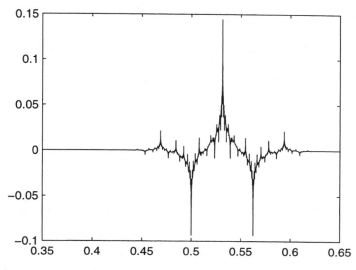

Fig. 5.7. Wavelet for CDF(2,2)

As stated above, these results are strictly correct only in a limiting sense, and they are not easy to establish. There is one further complication which we have omitted to state clearly. If one performs the procedure above with a 1 close to the start or end of the vector, then there will in general be some strange effects, depending on how the transform has been implemented. We refer to these as boundary effects. They depend on how one makes up for missing samples in computations near the start or end of a finite vector, the so-called boundary corrections, which will be considered in detail in Chap. 10. We have already mentioned zero padding as one of the correction methods. This is what we have used indirectly in plotting for example Fig. 5.7, where we have taken a vector with a one at place 24, and zeroes everywhere else.

If we try to interpret these results, then we can say that the direct transform resolves the signal into components of the shape given by the scaling function and the wavelet. More precisely, it is a superposition of these components, with weight according to the value of the entry in the transform, since the basic shapes were based on vectors with entry equal to 1. This is a generalization of the concrete computation at the end of Sect. 5.2.2.

Readers interested in a rigorous treatment of the interpretation of the transforms given here, and with the required mathematical background, are referred to the literature, for example the books by I. Daubechies [5], S. Mallat [16], and M. Vetterli-J. Kovačević [28]. Note that these books base their treatment of the wavelet transforms on the concepts of multiresolution analysis and filter theory. We have not yet discussed these concepts.

Exercises

Note that the exercises 5.4–5.7 require use of MATLAB and *Uvi_Wave*. Postpone solving them until after you have read Sect. 13.1. This section is independent of the following chapters.

5.1 Carry out the missing five inverse transforms needed to find the matrix (5.1).

5.2 Carry out the computations leading to (5.2).

5.3 Carry out computations similar to those leading to (5.1), and (5.1), in order to find the matrices $\mathbf{W}_a^{(1)}$ and $\mathbf{W}_s^{(1)}$ for the one scale Haar DWT applied to a signal of length 8.

5.4 Carry out the computations leading to Fig. 5.3, as explained in the text.

5.5 Carry out the computations leading to Fig. 5.5. Then try the same for some of the 61 basis vectors not plotted in this figure.

5.6 Plot the scaling function of CDF(2,2). In MATLAB you can do this using the functions `wspline` and `wavelet` from the *Uvi_Wave* toolbox.

5.7 Carry out the computations leading to Fig. 5.6, as explained in the text, and then those leading to Fig. 5.7.

6. Two Dimensional Transforms

In this chapter we will briefly show how the discrete wavelet transform can be applied to two dimensional signals, such as images. The 2D wavelet transform comes in two forms. One which consists of two 1D transforms, and one which is a true 2D transform. The first type is called separable, and the second nonseparable. We present some results and examples in the separable case, since it is a straightforward generalization of the results in the one dimensional case. At the end of the chapter we give an example of a nonseparable 2D DWT based on an adaptation of the lifting technique to the 2D case.

In this chapter we will focus solely on grey scale images. Such an image can be represented by a matrix, where each entry gives the grey scale value of the corresponding pixel. The purpose of this chapter is therefore to show how to apply the DWT to a matrix as opposed to a vector, as we did in previous chapters.

6.1 One Scale DWT in Two Dimensions

We use the notation $\mathbf{X} = \{x[m,n]\}$ to represent a matrix. As an example we have an 8×8 matrix

$$\mathbf{X} = \begin{bmatrix} x[1,1] & x[1,2] & \dots & x[1,8] \\ x[2,1] & x[2,2] & \dots & x[2,8] \\ \vdots & \vdots & \ddots & \vdots \\ x[8,1] & x[8,2] & \dots & x[8,8] \end{bmatrix}.$$

One way of applying the one dimensional technique to this matrix is by interpreting it as a one dimensional digital signal, simply by concatenating the rows as shown here

$$x[1,1], x[1,2], \cdots, x[1,8], x[2,1], x[2,2], \cdots, x[8,8] \ .$$

This yields a signal of length 64. The one dimensional discrete wavelet transform can then be applied to this signal. However, this is usually not a good approach, since there can be correlation between entries in neighboring rows. For example, there can be large areas of the image with the same grey scale value. These neighboring samples are typically not neighbors in the 1D signal obtained by concatenating the rows, and hence the transform may not detect the correlation.

Fortunately, there is a different way of applying the 1D transform to a matrix. We recall from Chap. 5 that we can represent the wavelet transform itself as a matrix, see in particular Exer. 5.3. This fact is also discussed in more details in Sect. 10.2. In the present context we take a matrix $\mathbf{W_a}$, which performs a one scale wavelet transformation, when applied to a column vector. To simplify the notation we have omitted the superscript (1). We apply this transform to the first column in the matrix $\mathbf{X}$,

$$\begin{bmatrix} y^c[1,1] \\ y^c[2,1] \\ \vdots \\ y^c[8,1] \end{bmatrix} = \begin{bmatrix} w[1,1] & w[1,2] & \dots & w[1,8] \\ w[2,1] & w[2,2] & \dots & w[2,8] \\ \vdots & \vdots & \ddots & \vdots \\ w[8,1] & w[8,2] & \dots & w[8,8] \end{bmatrix} \begin{bmatrix} x[1,1] \\ x[2,1] \\ \vdots \\ x[8,1] \end{bmatrix} .$$

The superscript 'c' is an abbreviation for 'column.' The same operation is performed on the remaining columns. But this is just ordinary matrix multiplication. We write the result as

$$\mathbf{Y}^c = \mathbf{W_a X} . \tag{6.1}$$

We then perform the same operation on the rows of $\mathbf{Y}^c$. This can be done by first transposing $\mathbf{Y}^c$, then multiplying it by $\mathbf{W_a}$, and finally transposing again. The transpose of a matrix $\mathbf{A}$ is denoted by $\mathbf{A}^\top$. Thus the result is

$$\mathbf{Y}^{c,r} = \left(\mathbf{W_a} (\mathbf{Y}^c)^\top \right)^\top = \mathbf{Y}^c \mathbf{W_a}^\top ,$$

by the usual rules for the transpose of a matrix. We can summarize these computations in the equation

$$\mathbf{Y}^{c,r} = \mathbf{W_a X W_a}^\top . \tag{6.2}$$

The superscripts show that we have first transformed columns, and then rows. But (6.2) shows that the same result is obtained by first transforming rows, and then columns, since matrix multiplication is associative, $(\mathbf{W_a X}) \mathbf{W_a}^\top = \mathbf{W_a} (\mathbf{X W_a}^\top)$.

We can find the inverse transform by using the rules of matrix computations. The result is

$$\mathbf{X} = \mathbf{W_a}^{-1} \mathbf{Y}^{c,r} (\mathbf{W_a}^{-1})^\top = \mathbf{W_s} \mathbf{Y}^{c,r} \mathbf{W_s}^\top . \tag{6.3}$$

Here $\mathbf{W_s} = \mathbf{W_a}^{-1}$ denotes the synthesis matrix, see Chap. 5, in particular Exer. 5.3. This gives a fairly simple method for finding the discrete wavelet transform of a two dimensional signal. As usual, we do not use matrix multiplication when implementing this transform. It is much more efficient to do two one dimensional transforms, implemented as lifting steps.

Thus in this section and the next we use the above definition of the separable DWT, and apply it to 2D signals. The properties derived in the previous chapters still hold, since we just use two 1D transforms. But there are also new properties related to the fact that we now have a 2D transform. In the following section we will discuss some of these properties through a number of examples.

6.2 Interpretation and Examples

We will again start with the Haar transform, in order to find out, how we can interpret the transformed image. We use the same ordering of the entries in the transform as in Table 3.1, dividing the transform coefficients into separate low and high pass parts. After a one scale Haar transform on both rows and columns, we end up with a matrix that naturally is interpreted as consisting of four submatrices, as shown in Fig. 6.1.

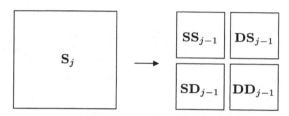

Fig. 6.1. Interpretation of the two dimensional DWT

The notation is consistent with the one used for one dimensional signals. The lower index j labels the size of the matrix. Thus S_j is a $2^j \times 2^j$ matrix. The submatrix SS_{j-1} of size $2^{j-1} \times 2^{j-1}$ consists of entries that contain means over both columns and rows. In the part SD_{j-1} we have computed means for the columns and differences for the rows. The two operations are reversed in the part DS_{j-1}. In the part DD_{j-1} we have computed differences for both rows and columns.

We use this one scale Haar-based transform as the building block for a multiresolution!two dimensional 2D DWT. We perform a one scale 2D transform, and then iterate on the averaged part SS_{j-1}, in order to get the next step. We will illustrate the process on some simple synthetic images. We start with the image given in Fig. 6.2.

We now perform a one scale 2D Haar transform on this image, and obtain the results shown in Fig. 6.3. The left hand plot shows the coefficients, and the right hand plot the inverse transform of each of the four blocks with the other three blocks equal to zero. The right hand plot is called the two dimensional multiresolution representation, since it is the 2D analogue of

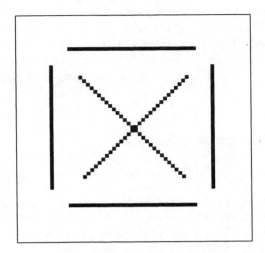

Fig. 6.2. A synthetic image

the multiresolution representations introduced in Chap. 4. We now change our one scale transform to the CDF(2,2) transform. Then we repeat the computations above. The result is shown in Fig. 6.4. Note that this transform averages over more pixels than the Haar transform, which is clearly seen in the **SS** component.

Let us explain in some detail how these figures have been obtained. The synthetic image in Fig. 6.2 is based on a 64 × 64 matrix. The black pixels correspond to entries with value 1. All other entries have value zero. The figure is shown with a border. The border is not part of the image, but is included in order to be able to distinguish the image from the white background. Borders are omitted in the following figures.

We have computed the Haar transform as described, and plotted the transform in Fig. 6.3 on the left hand side. Then we have carried out the multiresolution computation and plotted the result on the right hand side. Multiresolutions are computed in two dimensions in analogy with the one dimensional case. We select a component of the decomposition, and replace the other three components by zeroes. Then we compute the inverse transform. In this way we obtain the four parts of the picture on the right hand side of Fig. 6.3.

The grey scale—imp has been adjusted, such that white pixels correspond to the value −1 and black pixels to the value 1. The large medium grey areas correspond to the value 0. The same procedure is repeated in Fig. 6.4, using the CDF(2,2) transform.

The one step examples clearly show the averaging, and the emphasis of vertical, horizontal, and diagonal lines, respectively, in the four components.

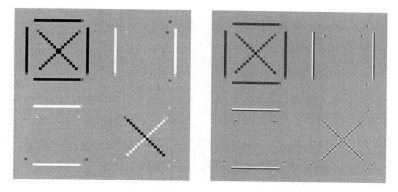

Fig. 6.3. One step 2D Haar transform, grey scale adjusted. Left hand plot shows coefficients, right hand plot the inverse transform of each block, the multiresolution representation

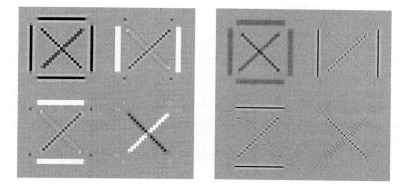

Fig. 6.4. One step 2D CDF(2,2) transform, grey scale adjusted. Left hand plot shows coefficients, right hand plot the inverse transform of each block, the multiresolution representation

We now do a two step CDF(2,2) transform. The result is shown in Fig. 6.5. This time we have only shown a plot of the coefficients. In order to be able to see details, the grey scale has been adjusted in each of the blocks of the transform, such that the largest value in a block corresponds to black and the smallest value in a block to white.

Let us next look at some more complicated synthetic images. All images have 128×128 pixels. On the left hand side in Fig. 6.6 we have taken an image with distinct vertical and horizontal lines. Using the Haar building block, and transforming over three scales, we get the result on the right hand side in Fig. 6.6. We see that the vertical and horizontal lines are clearly separated into the respective components of the transform. The diagonal components

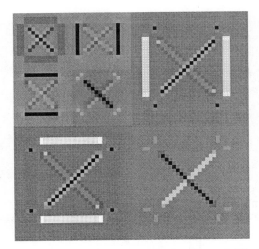

Fig. 6.5. CDF(2,2) transform, 2 scales, grey scale adjusted. Plot of the coefficients

contain little information on the line structure, as expected. In this figure the grey scale has again been adjusted in each of the ten blocks. The plot shows the coefficients.

We now take the image on the left hand side in Fig. 6.7, with a complicated structure. The Haar transform over three scales is also shown. Here the directional effects are much less pronounced, as we would expect from the original image. Again, the plot shows the coefficients.

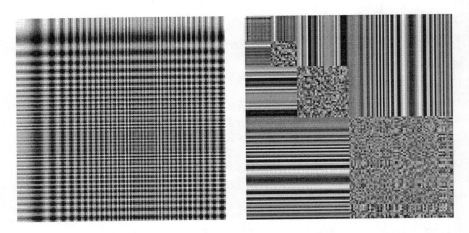

Fig. 6.6. Left: Synthetic image with vertical and horizontal lines. Right: Haar transform of image over 3 scales, grey scale adjusted. Plot of the coefficients

Fig. 6.7. Left: Synthetic image with complex structure. Right: Haar transform of image over 3 scales, grey scale adjusted. Plot of the coefficients

Finally, we try with the real image 'Lena' shown in the left hand image in Fig. 6.8. This image is often used in the context of image processing. In this case we have chosen a resolution of 256×256 pixels. The decomposition over two scales is shown on the right hand side, again with the grey scale adjusted within each block. The plot is of the coefficients.

In Fig. 6.9 we have set the averaged part of the transform equal to zero, keeping the six detail blocks, and then applied the inverse 2D transform over the two scales, in order to locate contours in the image. We have adjusted the grey scale to be able to see the details.

In this section we have experimented with separable 2D DWTs. As the examples have shown, there is a serious problem with separable transforms, since they single out horizontal, vertical, and diagonal structures in the given image. In a complicated image a small rotation of the original could change the features emphasized drastically.

6.3 A 2D Transform Based on Lifting

We describe briefly an approach leading to a nonseparable two dimensional discrete wavelet transform. It is based on the lifting ideas from Sect. 3.1. The starting point is the method we used to introduce the Haar transform, and also the CDF(2,2) transform, in Sect. 3.2, namely consideration of the nearest neighbors.

To avoid problems with the boundary, we look at an infinite image, where the pixels have been labeled by pairs of integers, such that the image is described by an infinite matrix $\mathbf{X} = \{x[m,n]\}_{(m,n) \in \mathbf{Z} \times \mathbf{Z}}$. The key concept is the nearest neighbor. Each entry $x[m,n]$ has four nearest neighbors, namely

Fig. 6.8. Left: Standard image Lena in 256 × 256 resolution. Right: CDF(2,2) transform over 2 scales of 'Lena.' Image is grey scale adjusted

Fig. 6.9. Reconstruction based on the detail parts in Fig. 6.8. Image is grey scale adjusted

the entries $x[m+1, n]$, $x[m-1, n]$, $x[m, n+1]$, and $x[m, n-1]$. This naturally leads to a division of all points in the plane with integer coordinates into two classes. This division is defined as follows. We select a point as our starting point, for example the origin $(0, 0)$, and color it black. This point has four nearest neighbors, which we assign the color white. Next we select one of the white points and color its four nearest neighbors black. One of them, our starting point, has already been assigned the color black. Continuing in this manner, we divide the whole lattice $\mathbf{Z} \times \mathbf{Z}$ into two classes of points, called black and white points. Each point belongs to exactly one class. We have illustrated the assignment in Fig. 6.10.

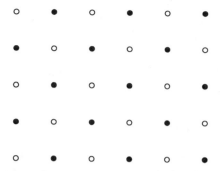

Fig. 6.10. Division of the integer lattice into black and white points

Comparing with the one dimensional case, then the black points correspond to entries in a one dimensional signal with odd indices, and the white points to those with even indices. We recall from Sect. 3.2 that the first step in the one dimensional case was to predict the value at an odd indexed entry, and then replace this entry with the difference, see (3.9). In the two dimensional case we do exactly the same. We start with the black points. Each black point value is replaced with the difference between the original value and the predicted value. This is done for all black points in the first step. In the second step we go through the white points and update the values here, based on the just computed values at the black points, see (3.10) in the one dimensional case. This is the one scale building block. To define a two dimensional discrete wavelet transform, we keep the computed values at the black points, and then use the lattice of white points as a new lattice, on which we perform the two operations in the building block. Notice that the white points in Fig. 6.10 constitute a square lattice, which is rotated 45° relative to the original integer lattice, and with a distance of $\sqrt{2}$ between nearest neighbors. This procedure is an 'in place' algorithm, since we work on just one matrix, successively computing new values and inserting them in the original matrix.

The transform is inverted exactly as in Sect. 3.2, namely by reversing the order of the operations and changing the signs.

Let us now make a specific choice of prediction and update procedures. For a given black point, located at (m, n), we predict that the value at this point should be the average of the nearest four neighbors. Thus we replace $x[m, n]$ by

$$x_\bullet[m, n] = x[m, n]$$
$$- \frac{1}{4} \left(x[m - 1, n] + x[m + 1, n] + x[m, n + 1] + x[m, n - 1] \right) .$$

We decide to use an update procedure, which preserves the average value. Thus the average of the computed values at the white points should equal half of the average over all the initial values. The factor one half comes from the fact that there are half as many white values as there are values in the original image. A simple computation shows that one can obtain this property by the following choice of update procedure

$$x_\circ[m,n] = x[m,n]$$
$$+ \frac{1}{8} \left(x_\bullet[m-1,n] + x_\bullet[m+1,n] + x_\bullet[m,n+1] + x_\bullet[m,n-1] \right) \ .$$

The discrete wavelet transform, defined this way, has the property that it does not exhibit the pronounced directional effects of the one defined in the first section. In the literature the lattice in Fig. 6.10 is called the quincunx lattice. It is possible to use other lattices, for example a hexagonal lattice, and other procedures for prediction and update.

Exercises

These exercises assume that the reader is familiar with the *Uvi_Wave* toolbox, as for example explained in Chap. 13

6.1 Go through the details of the examples in this section, using the *Uvi_Wave* toolbox in MATLAB.

6.2 Select some other images and perform computer experiments similar to those above.

6.3 (Difficult.) After you have read Chap. 11, try to implement the nonseparable two dimensional wavelet transform described above. Experiment with it and compare with some of the applications of the separable one based on the Haar building block, in particular with respect to directional effects.

7. Lifting and Filters I

Our discussion of the discrete wavelet transform has to this point been based on the time domain alone. We have represented and treated signals as sequences of sample points $\mathbf{x} = \{x[n]\}_{n \in \mathbf{Z}}$. The index $n \in \mathbf{Z}$ represents equidistant sampling times, given a choice of time scale. But we will get a better understanding of the transform, if we also look at it in the frequency domain. The frequency representation is obtained from the time domain representation using the Fourier transform. We will refer to standard texts [22, 23] for the necessary background, but we recall some of the definitions and results below.

In the previous chapters we have introduced the discrete wavelet transform using the lifting technique. In the literature, for example [5], the discrete wavelet transform is defined using filters. In this chapter we establish the connection between the definition using filters, and the lifting definition. We establish the connection both in the time domain, and in the frequency domain. Further results on the connection between filters and lifting are given in Chap. 12.

We note that in this chapter we only discuss the properties of a one scale discrete wavelet transform, as described in Chap. 3. We look at the properties of multiresolution transforms in Chap. 9.

Some of the results in this chapter are rather technical, but they are needed in the sequel for a complete discussion of the DWT. Consequently, in the following chapters we will often refer back to results obtained here. Note that it is possible to read the following chapters without having absorbed all the details in this chapter. One can return to this chapter, when the results are needed.

7.1 Fourier Series and the z-Transform

We first recall some standard results from signal analysis. A finite energy signal $\mathbf{x} = \{x[n]\}_{n \in \mathbf{Z}} \in \ell^2(\mathbf{Z})$ has an associated Fourier series,

$$X(e^{j\omega}) = \sum_{n \in \mathbf{Z}} x[n]e^{-jn\omega} . \tag{7.1}$$

The function $X(e^{j\omega})$, given by the sum of this series, is periodic with period 2π, as a function of the frequency variable ω. We denote the imaginary unit by j, i.e. $j^2 = -1$, as is customary in the engineering literature. We also adopt the custom that the upper case letter X denotes the representation of the sequence $\mathbf{x}$ in the frequency domain.

For each $\mathbf{x} \in \ell^2(\mathbf{Z})$ we have the following result, which is called Parseval's equation,

$$\frac{1}{2\pi} \int_0^{2\pi} \left| X(e^{j\omega}) \right|^2 d\omega = \sum_{n \in \mathbf{Z}} |x[n]|^2 . \tag{7.2}$$

This means that the energy of the signal can be computed in the frequency representation via this formula.

Conversely, given a 2π-periodic function $X(e^{j\omega})$, which satisfies

$$\int_0^{2\pi} \left| X(e^{j\omega}) \right|^2 d\omega < \infty , \tag{7.3}$$

then we can find a sequence $\mathbf{x} \in \ell^2(\mathbf{Z})$, such that (7.1) holds, by using

$$x[n] = \frac{1}{2\pi} \int_0^{2\pi} X(e^{j\omega}) e^{jn\omega} d\omega . \tag{7.4}$$

This is a consequence of the orthogonality relation for exponentials, expressed as

$$\frac{1}{2\pi} \int_0^{2\pi} e^{-jm\omega} e^{jn\omega} d\omega = \begin{cases} 0 & \text{if } m \neq n , \\ 1 & \text{if } m = n . \end{cases} \tag{7.5}$$

We note that there are technical details concerning the type of convergence for the series (7.1) (convergence in mean), which we have decided to omit here.

The z-transform is obtained from the Fourier representation (7.1) by substituting z for $e^{j\omega}$. Thus the z-transform of the sequence $\mathbf{x}$ is defined as

$$X(z) = \sum_{n \in \mathbf{Z}} x[n] z^{-n} . \tag{7.6}$$

Initially this equation is valid only for $z = e^{j\omega}$, $\omega \in \mathbf{R}$, or expressed in terms of the complex variable z, for z taking values on the unit circle. But in many cases we can extend the complex function $X(z)$ to a larger domain in the complex plane, and use techniques and results from complex analysis. In particular, for a finite signal (a signal with only finitely many non-zero entries) the transform $X(z)$ is a polynomial in z and z^{-1}. Thus it is defined in the whole complex plane, except perhaps at the origin, where it may have a pole.

The z-transform defined by (7.6) is linear. This means that the z-transform of $\mathbf{w} = \alpha\mathbf{x} + \beta\mathbf{y}$ is $W(z) = \alpha X(z) + \beta Y(z)$, where α and β are complex numbers.

Another property of the z-transform is that it transforms convolution of sequences into multiplication of the corresponding z-transforms. Let $\mathbf{x}$ and $\mathbf{y}$ be two sequences from $\ell^2(\mathbf{Z})$. The convolution of $\mathbf{x}$ and $\mathbf{y}$ is the sequence $\mathbf{w} = \mathbf{x} * \mathbf{y}$, defined by

$$w[n] = (\mathbf{x} * \mathbf{y})[n] = \sum_{k \in \mathbf{Z}} x[n - k]y[k] \ . \tag{7.7}$$

In the z-transform representation this relation becomes

$$W(z) = X(z)Y(z) \ . \tag{7.8}$$

Let us verify this result. We compute as follows, using the definitions. Sums extend over all integers.

$$
\begin{aligned}
W(z) &= \sum_n w[n]z^{-n} = \sum_n \left(\sum_k x[n - k]y[k] \right) z^{-n} \\
&= \sum_n \sum_k x[n - k]z^{-(n-k)}y[k]z^{-k} \\
&= \sum_k \left(\sum_n x[n - k]z^{-(n-k)} \right) y[k]z^{-k} \\
&= \sum_k X(z)y[k]z^{-k} = X(z)Y(z) \ .
\end{aligned}
$$

The relation (7.8) shows that we have

$$\mathbf{x} * \mathbf{y} = \mathbf{y} * \mathbf{x} \ . \tag{7.9}$$

This result can also be shown directly from the summation definition of convolution, using a change of variables.

In order for (7.7) to define a sequence $\mathbf{w}$ in $\ell^2(\mathbf{Z})$, and for (7.8) to define a function $W(z)$ satisfying (7.3), we need an additional condition on one of the sequences $\mathbf{x}$ and $\mathbf{y}$, or on one of the functions $X(z)$ and $Y(z)$. One possibility is to require that $X(e^{j\omega})$ is a bounded function for $\omega \in \mathbf{R}$. A stronger condition is to require that $\sum_n |x[n]| < \infty$. In many applications only finitely many entries in $\mathbf{x}$ are nonzero, and then both conditions are obviously satisfied.

Shifting a given signal one time unit left or right can also be implemented in the z-transform representation. Suppose $\mathbf{x} = \{x[n]\}$ is a given signal. Let $\mathbf{x}_{\text{left}} = \{x[n+1]\}$ denote the signal shifted one time unit to the left. Then it follows from the definition of the z-transform that we have

$$X_{\text{left}}(z) = \sum_n x[n+1]z^{-n} = \sum_n x[n+1]z^{-(n+1)+1} = zX(z) \ . \tag{7.10}$$

Analogously, let $\mathbf{x_{right}} = \{x[n-1]\}$ denote the signal shifted one time unit to the right. Then

$$X_{\text{right}}(z) = \sum_n x[n-1]z^{-n} = \sum_n x[n-1]z^{-(n-1)-1} = z^{-1}X(z) \ . \quad (7.11)$$

Two other operations needed below are down sampling and up sampling by two. Given a sequence $\mathbf{x}$, then this sequence down sampled by two, denoted by $\mathbf{x_{2\downarrow}}$, is defined in the time domain by

$$x_{2\downarrow}[n] = x[2n], \quad n \in \mathbf{Z} \ . \quad (7.12)$$

Described in words, this means that we delete the odd indexed entries in the given sequence, and then change the indexing. In the z-transform representation we find the following result (note that in the second sum the terms with k odd cancel),

$$\begin{aligned} X_{2\downarrow}(z) &= \sum_n x[2n]z^{-n} \\ &= \sum_k \frac{1}{2}\left(x[k](z^{1/2})^{-k} + x[k](-z^{1/2})^{-k} \right) \\ &= \frac{1}{2}\left(X(z^{1/2}) + X(-z^{1/2}) \right) \ . \end{aligned} \quad (7.13)$$

Given a sequence $\mathbf{y}$, the up sampling operation yields the sequence $\mathbf{y_{2\uparrow}}$, obtained in the time domain by

$$y_{2\uparrow}[n] = \begin{cases} 0 & \text{if } n \text{ is odd}, \\ y[n/2] & \text{if } n \text{ is even} . \end{cases} \quad (7.14)$$

This means that we interlace zeroes between the given samples, and then change the indexing. In the z-transform representation we find, after a change of summation variable from n to $k = n/2$,

$$Y_{2\uparrow}(z) = \sum_n y_{2\uparrow}[n]z^{-n} = \sum_k y[k]z^{-2k} = Y(z^2) \ . \quad (7.15)$$

As a final property we mention the uniqueness result for the z-transform representation. This can be stated as follows. If $X(z) = Y(z)$ for all z on the unit circle, then $x[n] = y[n]$ for all $n \in \mathbf{Z}$. This is a consequence of (7.4).

7.2 Lifting in the z-Transform Representation

We are now ready to show how to implement the one scale DWT, defined via the lifting technique, in the z-transform representation. The first step was to split a given signal into its even and odd components. In the z-transform representation this splitting is obtained by writing

$$X(z) = X_0(z^2) + z^{-1}X_1(z^2) , \tag{7.16}$$

where

$$X_0(z) = \sum_n x[2n]z^{-n} , \tag{7.17}$$

$$X_1(z) = \sum_n x[2n + 1]z^{-n} . \tag{7.18}$$

Using the results from the previous section, we see that X_0 is obtained from the original signal by down sampling by two. The component X_1 is obtained from the given signal by first shifting it one time unit to the left, and then down sample by two. Using the formulas (7.13) and (7.10) we thus have

$$X_0(z) = \frac{1}{2}\left(X(z^{1/2}) + X(-z^{1/2})\right) , \tag{7.19}$$

$$X_1(z) = \frac{z^{1/2}}{2}\left(X(z^{1/2}) - X(-z^{1/2})\right) . \tag{7.20}$$

We represent this decomposition by the diagram in Fig. 7.1. The diagram should be compared with the formulas (7.19) and (7.20), which show the result of the decomposition, expressed in terms of $X(z)$.

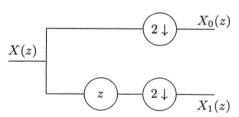

Fig. 7.1. Splitting in even and odd components

The inverse operation is obtained by reading equation (7.16) from right to left. The equation tells us that we can obtain $X(z)$ from $X_0(z)$ and $X_1(z)$ by first up sampling the two components by 2, then shifting $X_1(z^2)$ one time unit right (by multiplication by z^{-1}), and finally adding the two components. We represent this reconstruction by the diagram in Fig. 7.2.

Let us now see how we can implement the prediction step from Sect. 3.2 in the z-transform representation. The prediction technique was to form a linear combination of the even entries and then subtract the result from the odd entry under consideration. The linear combination was formed independently of the index of the odd sample under consideration, and based only on the

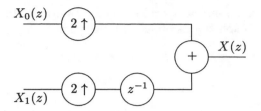

Fig. 7.2. Reconstructing the signal from even and odd components

relative location of the even entries. For example, in the CDF(2,2) transform the first step in (3.13) can be implemented as $X_1(z) - T(z)X_0(z)$, where $T(z) = \frac{1}{2}(1 + z)$. (Recall that $T(z)X_0(z)$ means the convolution $\mathbf{t} * \mathbf{x_0}$ in the time domain, which is exactly a linear combination of the even entries with weights $t[n]$.) Let us verify this result. First we multiply, using the definition (7.17), and then we change the summation variable in the second sum, to get

$$T(z)X_0(z) = \frac{1}{2}(1 + z)\sum_n x[2n]z^{-n}$$

$$= \frac{1}{2}\sum_n x[2n]z^{-n} + \frac{1}{2}\sum_n x[2n]z^{-n+1}$$

$$= \sum_n \frac{1}{2}(x[2n] + x[2n + 2])z^{-n} .$$

Thus we have

$$X_1(z) - T(z)X_0(z) = \sum_n \left(x[2n + 1] - \tfrac{1}{2}(x[2n] + x[2n + 2])\right)z^{-n} ,$$

which is exactly the z-transform representation of the right hand side of (3.13). The transition from $(X_0(z), X_1(z))$ to $(X_0(z), X_1(z) - T(z)X_0(z))$ can be described by matrix multiplication. We have

$$\begin{bmatrix} X_0(z) \\ X_1(z) - T(z)X_0(z) \end{bmatrix} = \begin{bmatrix} 1 & 0 \\ -T(z) & 1 \end{bmatrix} \begin{bmatrix} X_0(z) \\ X_1(z) \end{bmatrix} .$$

An entirely analogous computation (see Exer. 7.2) shows that if we define $S(z) = \frac{1}{4}(1 + z^{-1})$, then the update step in (3.14) is implemented in the z-transform representation as multiplication by the matrix

$$\begin{bmatrix} 1 & S(z) \\ 0 & 1 \end{bmatrix} .$$

The final normalization step in the various transforms in Chap. 3, as for example given in (3.31) and (3.30), can all be implemented by multiplication by a matrix of the form

$$\begin{bmatrix} K & 0 \\ 0 & K^{-1} \end{bmatrix},$$

where $K > 0$ is a constant. Note that this particular form depends on an overall normalization of the transform, as explained in connection with Theorem 7.3.1.

It is a rather surprising fact that the same simple structure of the lifting steps used above applies to the general case. In the general case a prediction step is always given by multiplication by a matrix of the form

$$\mathbf{P}(z) = \begin{bmatrix} 1 & 0 \\ -T(z) & 1 \end{bmatrix},$$

and an update step by multiplication by a matrix of the form

$$\mathbf{U}(z) = \begin{bmatrix} 1 & S(z) \\ 0 & 1 \end{bmatrix}.$$

Here $T(z)$ and $S(z)$ are both polynomials in z and z^{-1}. Such polynomials are called Laurent polynomials.

The general one scale DWT described in Chap. 3, with the normalization step included, is then in the z-transform representation given as a matrix product (see also Fig. 3.5)

$$\mathbf{H}(z) = \begin{bmatrix} K & 0 \\ 0 & K^{-1} \end{bmatrix} \begin{bmatrix} 1 & S_N(z) \\ 0 & 1 \end{bmatrix} \begin{bmatrix} 1 & 0 \\ -T_N(z) & 1 \end{bmatrix} \cdots \begin{bmatrix} 1 & S_1(z) \\ 0 & 1 \end{bmatrix} \begin{bmatrix} 1 & 0 \\ -T_1(z) & 1 \end{bmatrix}. \quad (7.21)$$

The order of the factors is determined by the order in which we apply the various steps. First a prediction step, then an update step, perhaps repeated N times, and then finally the normalization step. Note that matrix multiplication is non-commutative, i.e. $\mathbf{U}(z)\mathbf{P}(z) \neq \mathbf{P}(z)\mathbf{U}(z)$ in general.

An important property of the DWT implemented via lifting steps was the invertibility of the transform, as illustrated for example in Fig. 3.4. It is easy to verify that we have

$$\mathbf{P}(z)^{-1} = \begin{bmatrix} 1 & 0 \\ T(z) & 1 \end{bmatrix} \quad \text{and} \quad \mathbf{U}(z)^{-1} = \begin{bmatrix} 1 & -S(z) \\ 0 & 1 \end{bmatrix}. \quad (7.22)$$

Since

$$\left(\mathbf{U}(z)\mathbf{P}(z) \right)^{-1} = \mathbf{P}(z)^{-1}\mathbf{U}(z)^{-1}$$

by the usual rules of matrix multiplication, we have that the matrix $\mathbf{H}(z)$ in (7.21) is invertible, and its inverse, denoted by $\mathbf{G}(z)$, is given by

$$\mathbf{G}(z) = \begin{bmatrix} 1 & 0 \\ T_1(z) & 1 \end{bmatrix} \begin{bmatrix} 1 & -S_1(z) \\ 0 & 1 \end{bmatrix} \cdots \begin{bmatrix} 1 & 0 \\ T_N(z) & 1 \end{bmatrix} \begin{bmatrix} 1 & -S_N(z) \\ 0 & 1 \end{bmatrix} \begin{bmatrix} K^{-1} & 0 \\ 0 & K \end{bmatrix}. \quad (7.23)$$

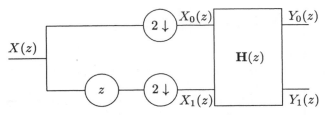

Fig. 7.3. One scale DWT in the z-representation, based on (7.21)

Multiplying all the matrices in the product defining $\mathbf{H}(z)$ in (7.21), we get a matrix with entries, which are Laurent polynomials. We use the notation

$$\mathbf{H}(z) = \begin{bmatrix} H_{00}(z) & H_{01}(z) \\ H_{10}(z) & H_{11}(z) \end{bmatrix}, \qquad (7.24)$$

for such a general matrix. We can then represent the implementation of the complete one scale DWT in the z-transform representation by the diagram in Fig. 7.3. Written in matrix notation the DWT is given as

$$\begin{bmatrix} Y_0(z) \\ Y_1(z) \end{bmatrix} = \begin{bmatrix} H_{00}(z) & H_{01}(z) \\ H_{10}(z) & H_{11}(z) \end{bmatrix} \begin{bmatrix} X_0(z) \\ X_1(z) \end{bmatrix}. \qquad (7.25)$$

This representation of a two channel filter bank, without any reference to lifting, is in the signal analysis literature called the *polyphase* representation, see [28].

The analogous diagram and matrix representation for the inversion are easily found, and are omitted here.

In the factored form it was easy to see that the matrix $\mathbf{H}(z)$ was invertible. In the form (7.24) invertibility may not be so obvious. But, just as for ordinary matrices, one can here use the determinant to decide whether a given matrix $\mathbf{H}(z)$ is invertible. The following proposition demonstrates how this is done.

Proposition 7.2.1. *A matrix* (7.24), *whose entries are Laurent polynomials, is invertible, if and only if its determinant is a monomial.*

Proof. The determinant is defined by

$$d(z) = \det \mathbf{H}(z) = H_{00}(z)H_{11}(z) - H_{01}(z)H_{10}(z), \qquad (7.26)$$

and it is a Laurent polynomial. A straightforward multiplication of matrices shows that

$$\begin{bmatrix} H_{00}(z) & H_{01}(z) \\ H_{10}(z) & H_{11}(z) \end{bmatrix} \begin{bmatrix} H_{11}(z) & -H_{01}(z) \\ -H_{10}(z) & H_{00}(z) \end{bmatrix} = d(z) \begin{bmatrix} 1 & 0 \\ 0 & 1 \end{bmatrix}, \qquad (7.27)$$

just like in ordinary linear algebra. This equation shows that the matrix $\mathbf{H}(z)$ is invertible, if and only if $d(z)$ is invertible. It also gives an explicit formula for the inverse matrix.

A Laurent polynomial is invertible, if and only if it is a monomial, i.e. it is of the form cz^k for some nonzero complex number c and some integer k. Let us verify this result. Let $p(z) = cz^k$. Then the inverse is $q(z) = c^{-1}z^{-k}$. Conversely, let

$$p(z) = a_{n_1} z^{n_1} + a_{n_1+1} z^{n_1+1} + \cdots + a_{n_2} z^{n_2} \, ,$$
$$q(z) = b_{m_1} z^{m_1} + b_{m_1+1} z^{m_1+1} + \cdots + b_{m_2} z^{m_2} \, ,$$

be two Laurent polynomials. Here $n_1 \leq n_2$ and $m_1 \leq m_2$ are integers, and the a_i and b_i are complex numbers. We assume that $a_{n_1} \neq 0$, $a_{n_2} \neq 0$, $b_{m_1} \neq 0$, and $b_{m_2} \neq 0$. Suppose now that $p(z)q(z) = 1$ for all nonzero complex numbers z. If $n_1 = n_2$ and $m_1 = m_2$, both p and q are monomials, and the result has been shown. So assume for example $n_1 < n_2$. We first multiply the two polynomials.

$$p(z)q(z) = a_{n_1} b_{m_1} z^{n_1+m_1} + \cdots + a_{n_2} b_{m_2} z^{n_2+m_2} \, .$$

We have assumed that $p(z)q(z) = 1$. But in the product we have at least two different powers of z with nonzero coefficients, namely $n_1 + m_1$ and $n_2 + m_2$, which is a contradiction. This shows the result.

The formula (7.27) shows that the usual formula for the inverse of a 2×2 matrix also holds in this case.

7.3 Two Channel Filter Banks with Perfect Reconstruction

Let us now present an approach to the one scale DWT, which is common in the signal analysis literature, see [28]. It is based on the concept of a two channel filter bank. First we need to introduce the concept of a filter. A filter is a linear map, which maps a signal with finite energy into another signal with finite energy. In the time domain it is given by convolution by a vector $\mathbf{h}$. To preserve the finite energy property we need to assume that the z-transform of this vector, $H(z)$, is bounded on the unit circle. Convolution by $\mathbf{h}$ is called filtering by $\mathbf{h}$. In the time domain it is given by $\mathbf{h} * \mathbf{x}$, and in the frequency domain by $H(z)X(z)$. The vector $\mathbf{h}$ is called the impulse response (IR) of the filter, (or sometimes the filter taps), and $H(e^{j\omega})$ the transfer function (or sometimes the frequency response). If $\mathbf{h}$ is a finite sequence, then $\mathbf{h}$ is called a FIR filter. Here FIR stands for finite impulse response. An infinite $\mathbf{h}$ is then called an IIR filter. We only consider FIR filters in this book. We also only consider filters with real coefficients. Further details on filtering can be found in any book on signal analysis, for example in [22, 23].

A two channel filter bank starts with two analysis filters, denoted by $\mathbf{h}_0$ and $\mathbf{h}_1$ here[1], and two synthesis filters, denoted by $\mathbf{g}_0$ and $\mathbf{g}_1$. All four filters

---

[1] Note that from Chap. 9 and onwards this notation is changed

are assumed to be FIR filters. Usually the filters with index 0 are chosen to be *low pass filters*, and the filters with index 1 to be *high pass filters*. In the usual terminology a low pass filter is a filter which is close to 1 for $|\omega| \leq \pi/2$, and close to 0 for $\pi/2 \leq |\omega| \leq \pi$. Similarly a high pass filter is close to 1 for $\pi/2 \leq |\omega| \leq \pi$, and close to 0 for $|\omega| \leq \pi/2$. In our case the value 1 has to be replaced by $\sqrt{2}$, due to the manner in which we have chosen to normalize the transform. We keep this normalization to facilitate comparison with the literature.

The analysis and synthesis parts of the filter bank are shown in Fig. 7.4. For the moment we consider the filtering scheme in the z-transform representation. Later we will also look at it in the time domain. The analysis part transforms the input $X(z)$ to the output pair $Y_0(z)$, $Y_1(z)$. The synthesis part then transforms this pair to the output $\widetilde{X}(z)$. The filtering scheme is said to have the *perfect reconstruction* property, if $X(z) = \widetilde{X}(z)$ for all possible (finite energy) $X(z)$.

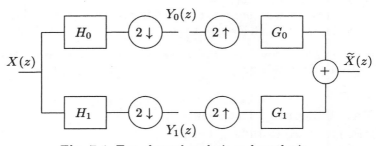

Fig. 7.4. Two channel analysis and synthesis

We first analyze which conditions are needed on the four filters, in order to obtain the perfect reconstruction property. We perform this analysis in the z-transform representation. Filtering by $\mathbf{h_0}$ transforms $X(z)$ to $H_0(z)X(z)$, and we then use (7.13) to down sample by two. Thus we have

$$Y_0(z) = \frac{1}{2}\left(H_0(z^{1/2})X(z^{1/2}) + H_0(-z^{1/2})X(-z^{1/2})\right), \qquad (7.28)$$

$$Y_1(z) = \frac{1}{2}\left(H_1(z^{1/2})X(z^{1/2}) + H_1(-z^{1/2})X(-z^{1/2})\right). \qquad (7.29)$$

Up sampling by two (see (7.15)), followed by filtering by the G-filters, and addition of the results, leads to a reconstructed signal

$$\widetilde{X}(z) = G_0(z)Y_0(z^2) + G_1(z)Y_1(z^2). \qquad (7.30)$$

Perfect reconstruction means that $\widetilde{X}(z) = X(z)$. We combine the above expressions and then regroup terms to get

$$\tilde{X}(z) = \frac{1}{2}[G_0(z)H_0(z) + G_1(z)H_1(z)]X(z)$$
$$+ \frac{1}{2}[G_0(z)H_0(-z) + G_1(z)H_1(-z)]X(-z) .$$

The condition $\tilde{X}(z) = X(z)$ will then follow from the conditions

$$G_0(z)H_0(z) + G_1(z)H_1(z) = 2 , \qquad (7.31)$$
$$G_0(z)H_0(-z) + G_1(z)H_1(-z) = 0 . \qquad (7.32)$$

The converse obviously also holds. These conditions mean that the four filters cannot be chosen independently, if we want to have perfect reconstruction.

Let us analyze the consequences of (7.31) and (7.32). We write them in matrix form

$$\begin{bmatrix} H_0(z) & H_1(z) \\ H_0(-z) & H_1(-z) \end{bmatrix} \begin{bmatrix} G_0(z) \\ G_1(z) \end{bmatrix} = \begin{bmatrix} 2 \\ 0 \end{bmatrix} . \qquad (7.33)$$

In order to solve this equation with respect to $G_0(z)$ and $G_1(z)$ we need the matrix to be invertible. Let us denote its determinant by $d(z)$. Since we assume that all filters are FIR filters, $d(z)$ is a Laurent polynomial. To be invertible, it has to be a monomial, as shown in Proposition 7.2.1. This determinant satisfies $d(-z) = -d(z)$, as the following computations show.

$$d(-z) = H_0(-z)H_1(-(-z)) - H_0(-(-(z)))H_1(-z)$$
$$= -\big(H_0(z)H_1(-z) - H_0(-z)H_1(z)\big)$$
$$= -d(z) .$$

This means that the monomial $d(z)$ has to be an odd integer power of z, so we can write it as

$$d(z) = 2c^{-1}z^{-2k-1} \qquad (7.34)$$

for some integer k and some nonzero constant c. Using Cramer's rule to solve (7.33) we get

$$G_0(z) = \frac{\begin{vmatrix} 2 & H_1(z) \\ 0 & H_1(-z) \end{vmatrix}}{d(z)} = cz^{2k+1}H_1(-z) , \qquad (7.35)$$

$$G_1(z) = \frac{\begin{vmatrix} H_0(z) & 2 \\ H_0(-z) & 0 \end{vmatrix}}{d(z)} = -cz^{2k+1}H_0(-z) . \qquad (7.36)$$

These equations show that we can choose either the H-filter pair or the G-filter pair. We will assume that we have filters H_0 and H_1, subject to the condition that

$$d(z) = H_0(z)H_1(-z) - H_0(-z)H_1(z) = cz^{2k+1} \qquad (7.37)$$

for some integer k and nonzero constant c. Then G_0 and G_1 are determined by the equations (7.35) and (7.36), which means that they are unique up to a scaling factor and an odd shift in time.

In the usual definition of the DWT the starting point is a two channel filter bank with the perfect reconstruction property. The analysis part is then used to define the direct one scale DWT (the building block from Sect. 3.2), and the synthesis part is used for reconstruction.

It is an important result that the filtering approach, and the one based on lifting, actually are identical. This means that they are just two different ways of describing the same transformation from $X(z)$ to $Y_0(z)$, $Y_1(z)$. We will now start explaining this equivalence. Part of the explanation, in particular the proof of the equivalence, will be postponed to Chap. 12.

The first step is to show that the analysis step in Fig. 7.4 is equivalent to the analysis step summarized in Fig. 7.3 and in (7.25). Thus we want to find the equations relating the coefficients in the matrix (7.24) and the filters. The analysis step by both methods should yield the same result. To avoid the square root terms we compare the results after up sampling by two. We start with the equality

$$\begin{bmatrix} Y_0(z^2) \\ Y_1(z^2) \end{bmatrix} = \mathbf{H}(z^2) \begin{bmatrix} \frac{1}{2}(X(z) + X(-z)) \\ \frac{z}{2}(X(z) - X(-z)) \end{bmatrix},$$

where Y_0 and Y_1 are obtained from the filter bank approach, see (7.28) and (7.29), and the right hand side from the lifting approach (in the polyphase form), with $\mathbf{H}$ from (7.24). We have also inserted the (up sampled) expressions (7.19) and (7.20) on the right hand side. The first equation can then be written as

$$\frac{1}{2}\big(H_0(z)X(z) + H_0(-z)X(-z)\big) =$$
$$H_{00}(z^2)\frac{1}{2}\big(X(z) + X(-z)\big) + H_{01}(z^2)\frac{z}{2}\big(X(z) - X(-z)\big).$$

This leads to the relation

$$H_0(z) = H_{00}(z^2) + zH_{01}(z^2).$$

Note the similarity to (7.16). The relation for H_1 is found analogously, and then the relations for G_0 and G_1 can be found using the perfect reconstruction conditions (7.35) and (7.36) in the two cases. The relations are summarized here.

$$H_0(z) = H_{00}(z^2) + zH_{01}(z^2), \qquad (7.38)$$
$$H_1(z) = H_{10}(z^2) + zH_{11}(z^2), \qquad (7.39)$$
$$G_0(z) = G_{00}(z^2) + z^{-1}G_{01}(z^2), \qquad (7.40)$$
$$G_1(z) = G_{10}(z^2) + z^{-1}G_{11}(z^2). \qquad (7.41)$$

Note the difference in the decomposition of the H-filters and the G-filters. Thus in the polyphase representation we use

$$\mathbf{H}(z) = \begin{bmatrix} H_{00}(z) & H_{01}(z) \\ H_{10}(z) & H_{11}(z) \end{bmatrix}, \tag{7.42}$$

$$\mathbf{G}(z) = \mathbf{H}(z)^{-1} = \begin{bmatrix} G_{00}(z) & G_{10}(z) \\ G_{01}(z) & G_{11}(z) \end{bmatrix}. \tag{7.43}$$

Note the placement of entries in $\mathbf{G}(z)$, which differs from the usual notation for matrices. The requirement of perfect reconstruction in the polyphase formulation was the requirement that $\mathbf{G}(z)$ should be the inverse of $\mathbf{H}(z)$. It is possible to verify that invertibility of $\mathbf{H}(z)$ is equivalent with the perfect reconstruction property for the filter bank, see Exer. 7.4.

If we start with a filter bank, then we can easily define $\mathbf{H}(z)$ using (7.38) and (7.39). But to get from $\mathbf{H}(z)$ to the lifting implementation, we need to factor $\mathbf{H}(z)$ into lifting steps (and a final normalization step), as in (7.21). The remarkable result is that this is always possible. This result was obtained by I. Daubechies and W. Sweldens in the paper [7]. In the general case $\det \mathbf{H}(z) = cz^{2k+1}$, whereas in (7.21) the determinant obviously is equal to 1. One can always get the determinant equal to one by scaling and an odd shift in time, so we state the result in this case.

Theorem 7.3.1 (Daubechies and Sweldens). *Assume that $\mathbf{H}(z)$ is a 2×2 matrix of Laurent polynomials, normalized to $\det \mathbf{H}(z) = 1$. Then there exists a constant $K \neq 0$ and Laurent polynomials $S_1(z), \ldots, S_N(z)$, $T_1(z), \ldots, T_N(z)$, such that*

$$\mathbf{H}(z) = \begin{bmatrix} K & 0 \\ 0 & K^{-1} \end{bmatrix} \begin{bmatrix} 1 & S_N(z) \\ 0 & 1 \end{bmatrix} \begin{bmatrix} 1 & 0 \\ -T_N(z) & 1 \end{bmatrix} \cdots \begin{bmatrix} 1 & S_1(z) \\ 0 & 1 \end{bmatrix} \begin{bmatrix} 1 & 0 \\ -T_1(z) & 1 \end{bmatrix}.$$

The proof of this theorem is constructive. It gives an algorithm for finding the Laurent polynomials $S_1(z), \ldots, S_N(z), T_1(z), \ldots, T_N(z)$ in the factorization. It is important to note that the factorization is *not unique*. Once we have a factorization, we can translate it into lifting steps. We will give some examples of this, together with a detailed proof of the theorem, in Chap. 12.

The advantage of the lifting approach, compared to the filter approach, is that it is very easy to find perfect reconstruction filters H_0, H_1, G_0, and G_1. It is just a matter of multiplying the lifting steps as in (7.21), and then assemble the filters according to the equations (7.38)–(7.41). In Sect. 7.8 we give some examples.

This approach should be contrasted with the traditional signal analysis approach, where one tries to find (approximate numerical) solutions to the equations (7.31) and (7.32), using for example spectral factorization. The weakness in constructing a transform based solely on the lifting technique is that it is based entirely on considerations in the time domain. Sometimes it is desirable to design filters with certain properties in the frequency domain, and once filters have been constructed in the frequency domain, we can use the constructive proof of the theorem to derive a lifting implementation, as explained in detail in Chap. 12. Another weakness of the lifting approach should be mentioned. The numerical stability of transforms defined using lifting can be difficult to analyze. We will give some further remarks on this problem in Sect. 14.2.

7.4 Orthonormal and Biorthogonal Bases

The two channel filter banks with perfect reconstruction discussed in the previous section can also be implemented in the time domain using convolution by filters. (Actually, this is the way the DWT is implemented in the *Uvi_Wave* toolbox (see Chap. 11), and in most other wavelet software packages.) This leads to an interpretation in terms of biorthogonal or orthonormal bases in $\ell^2(\mathbf{Z})$. In this section we define these bases and give a few results on them.

In $\ell^2(\mathbf{Z})$ one defines an inner product for any $\mathbf{x}, \mathbf{y} \in \ell^2(\mathbf{Z})$ by

$$\langle \mathbf{x}, \mathbf{y} \rangle = \sum_{n \in \mathbf{Z}} \overline{x}_n y_n \,. \tag{7.44}$$

Here $\overline{x}_n$ denotes complex conjugation. The inner product is connected to the norm via the equation $\|\mathbf{x}\| = \langle \mathbf{x}, \mathbf{x} \rangle^{1/2}$. Two vectors $\mathbf{x}, \mathbf{y} \in \ell^2(\mathbf{Z})$ are said to be *orthogonal*, if $\langle \mathbf{x}, \mathbf{y} \rangle = 0$.

At this point we need to introduce the Kronecker delta. It is the sequence defined by

$$\delta[n] = \begin{cases} 1 & \text{if } n = 0 \,, \\ 0 & \text{if } n \neq 0 \,. \end{cases} \tag{7.45}$$

Sometimes $\delta[n]$ is also viewed as a function of n.

Let $\mathbf{e}_n$, $n \in \mathbf{Z}$, be a sequence of vectors from $\ell^2(\mathbf{Z})$. It is said to be an *orthonormal* basis for $\ell^2(\mathbf{Z})$, if the following two properties hold.

orthonormality   For all $m, n \in \mathbf{Z}$ we have $\langle \mathbf{e}_m, \mathbf{e}_n \rangle = \delta[m - n]$.

completeness   If a vector $\mathbf{x} \in \ell^2(\mathbf{Z})$ satisfies $\langle \mathbf{x}, \mathbf{e}_n \rangle = 0$ for all $n \in \mathbf{Z}$, then $\mathbf{x} = \mathbf{0}$, where $\mathbf{0}$ denotes the zero vector.

An orthonormal basis has many nice properties. For example, it gives a representation for every vector $\mathbf{x} \in \ell^2(\mathbf{Z})$, as follows

$$\mathbf{x} = \sum_n \langle \mathbf{e}_n, \mathbf{x} \rangle \mathbf{e_n} \,. \tag{7.46}$$

For the inner product we have

$$\langle \mathbf{x}, \mathbf{y} \rangle = \sum_n \langle \mathbf{x}, \mathbf{e}_n \rangle \langle \mathbf{e}_n, \mathbf{y} \rangle \, , \tag{7.47}$$

which gives the following expression for the norm squared (the energy in the signal)

$$\|\mathbf{x}\|^2 = \sum_n |\langle \mathbf{x}, \mathbf{e}_n \rangle|^2 \, .$$

An example of an orthonormal basis is the so-called canonical basis, which is defined by

$$\mathbf{e}_n[k] = \delta[k - n] \, . \tag{7.48}$$

This means that the vector $\mathbf{e}_n$ has a one as the n'th entry and zeroes everywhere else. It is an exercise to verify that this definition actually gives an orthonormal basis, see Exer. 7.5.

Sometimes the requirements in the definition of the orthonormal basis are too restrictive for a particular purpose. In this case biorthogonal bases can often be used instead. They are defined as follows. Two sequences $\mathbf{f}_n$, $n \in \mathbf{Z}$, and $\tilde{\mathbf{f}}_m$, $m \in \mathbf{Z}$, are said to constitute a *biorthogonal* pair of bases for $\ell^2(\mathbf{Z})$, if the following properties are satisfied.

biorthogonality   For all $m, n \in \mathbf{Z}$ we have $\langle \mathbf{f}_m, \tilde{\mathbf{f}}_n \rangle = \delta[m - n]$.

stability   There exist positive constants A, B, $\tilde{A}$, and $\tilde{B}$, such that for all vectors $\mathbf{x} \in \ell^2(\mathbf{Z})$ we have

$$A\|\mathbf{x}\|^2 \le \sum_n |\langle \mathbf{f}_n, \mathbf{x} \rangle|^2 \le B\|\mathbf{x}\|^2 \, , \tag{7.49}$$

$$\tilde{A}\|\mathbf{x}\|^2 \le \sum_n |\langle \tilde{\mathbf{f}}_n, \mathbf{x} \rangle|^2 \le \tilde{B}\|\mathbf{x}\|^2 \, . \tag{7.50}$$

The expansion in (7.46) is now replaced by the following two expansions

$$\mathbf{x} = \sum_n \langle \mathbf{f}_n, \mathbf{x} \rangle \tilde{\mathbf{f}}_n \, ,$$

$$\mathbf{x} = \sum_n \langle \tilde{\mathbf{f}}_n, \mathbf{x} \rangle \mathbf{f}_n \, ,$$

and the expansion of the inner product in (7.47) by

$$\langle \mathbf{x}, \mathbf{y} \rangle = \sum_n \langle \mathbf{x}, \mathbf{f}_n \rangle \langle \tilde{\mathbf{f}}_n, \mathbf{y} \rangle \, .$$

Comparing the definitions, we see that an orthonormal basis is a special case of a biorthogonal basis pair, namely one satisfying $\tilde{\mathbf{f}}_n = \mathbf{f}_n$ for all $n \in \mathbf{Z}$. The completeness property comes from the stability property, since $\langle \mathbf{x}, \mathbf{f}_n \rangle = 0$ for all n and (7.49) imply that $\mathbf{x} = \mathbf{0}$.

7.5 Two Channel Filter Banks in the Time Domain

Let us first look at the realization of the two channel filter bank in the time domain. Note that this is the implementation most often used, since real signals are given as sequences of samples.

As Fig. 7.4 shows, in the time domain we filter (using convolution with h_0 and h_1), and then down sample. Obviously the two steps should be combined, to avoid computing terms not needed. The result is

$$y_0[k] = \sum_n h_0[2k-n]x[n] = \sum_n h_0[n]x[2k-n] , \qquad (7.51)$$

$$y_1[k] = \sum_n h_1[2k-n]x[n] = \sum_n h_1[n]x[2k-n] . \qquad (7.52)$$

We have shown both forms of the convolution, see (7.9).

To get formulas for the inverse transform we look again at Fig. 7.4. In the time domain we first up sample y_0 and y_1, and then convolute with g_0 and g_1, respectively. Finally the two results are added. Again it is obvious that all three operations should be combined. The result is

$$x[k] = \sum_n \left(g_0[k-2n]y_0[n] + g_1[k-2n]y_1[n] \right) . \qquad (7.53)$$

This formulation avoids explicit introduction of zero samples in the up sampled signals. They are eliminated by changing to the variable $n/2$ in the summation defining the convolution.

We now look at the perfect reconstruction property of a two channel filter bank in the time domain. Thus we assume that we have four filters h_0, h_1, g_0, and g_1, such that the associated filter bank has the perfect reconstruction property. In particular, we know that equations (7.31),(7.32), (7.35), and (7.36) all hold for the z-transforms.

We have to translate these conditions into conditions in the time domain on the filter coefficients. These conditions are obtained through a rather lengthy series of straightforward computations. We start with the four equations mentioned above. We then derive four new equations below.

Using (7.35) and (7.36) we get

$$G_1(z)H_1(z) = G_0(-z)H_0(-z) . \qquad (7.54)$$

Using (7.31) and (7.54) we get the first equation

$$G_0(z)H_0(z) + G_0(-z)H_0(-z) = 2 . \qquad (7.55)$$

We observe that $H_0(-z)$ is the z-transform of the sequence $\{(-1)^n h_0[n]\}$, and then we note that the constant function 2 is the z-transform of the sequence $\{2\delta[n]\}$. Finally we use the uniqueness property of the z-transform and the z-transform representation of convolution to write (7.55) as

$$\sum_k g_0[k]h_0[n-k] + (-1)^n \sum_k g_0[k]h_0[n-k] = 2\delta[n] .$$

Now for n odd the two terms cancel, and for n even they are equal. Thus we find, replacing n by $2n$ (note that $\delta[2n] = \delta[n]$ by the definition),

$$\sum_k g_0[k]h_0[2n-k] = \delta[n] \quad \text{for all } n \in \mathbf{Z} . \tag{7.56}$$

We use (7.54) once more in (7.31), this time with z replaced by $-z$, to get the second equation

$$G_1(z)H_1(z) + G_1(-z)H_1(-z) = 2 . \tag{7.57}$$

A computation identical to the one leading to (7.56) yields

$$\sum_k g_1[k]h_1[2n-k] = \delta[n] \quad \text{for all } n \in \mathbf{Z} . \tag{7.58}$$

Using (7.35) we get

$$G_0(z)H_1(z) = -c^{-1}z^{-2k-1}G_0(z)G_0(-z) ,$$

and then replacing z by $-z$

$$G_0(-z)H_1(-z) = c^{-1}z^{-2k-1}G_0(z)G_0(-z) .$$

Adding these expressions we get the third equation

$$G_0(z)H_1(z) + G_0(-z)H_1(-z) = 0 . \tag{7.59}$$

We can translate this equation into the time domain as above. The result is

$$\sum_k g_0[k]h_1[2n-k] = 0 \quad \text{for all } n \in \mathbf{Z} . \tag{7.60}$$

Using (7.36) we get

$$G_1(z)H_0(z) = -cz^{2k+1}H_0(z)H_0(-z)$$

and

$$G_1(-z)H_0(-z) = cz^{2k+1}H_0(z)H_0(-z) .$$

Adding these yields the fourth equation

$$G_1(z)H_0(z) + G_1(-z)H_0(-z) = 0 . \tag{7.61}$$

As above this leads to

$$\sum_k g_1[k]h_0[2n - k] = 0 \quad \text{for all } n \in \mathbf{Z} \ . \tag{7.62}$$

Thus we have shown that the perfect reconstruction property of the filter bank leads to the four equations (7.56), (7.58), (7.60), and (7.62). It turns out that these four equations have a natural interpretation as the first biorthogonality condition in the definition of a biorthogonal basis pair in $\ell^2(\mathbf{Z})$. This can be seen in the following way. We define the sequences of vectors $\{\mathbf{f}_i\}$ and $\{\tilde{\mathbf{f}}_i\}$ by

$$f_{2n}[k] = h_0[2n - k], \quad f_{2n+1}[k] = h_1[2n - k] \ , \tag{7.63}$$

$$\tilde{f}_{2n}[k] = g_0[k - 2n], \quad \tilde{f}_{2n+1}[k] = g_1[k - 2n] \ , \tag{7.64}$$

for all $n \in \mathbf{Z}$. Since we assume that the four filters are FIR filters, these vectors all belong to $\ell^2(\mathbf{Z})$. The four relations above then lead to the following properties of these vectors, where we also use the definition of the inner product and the assumption of real filter coefficients.

$$\langle \tilde{\mathbf{f}}_0, \mathbf{f}_{2n} \rangle = \delta[n], \quad \langle \tilde{\mathbf{f}}_0, \mathbf{f}_{2n+1} \rangle = 0 \ ,$$

$$\langle \tilde{\mathbf{f}}_1, \mathbf{f}_{2n} \rangle = 0, \quad \langle \tilde{\mathbf{f}}_1, \mathbf{f}_{2n+1} \rangle = \delta[n] \ ,$$

for all $n \in \mathbf{Z}$. They lead to the biorthogonality condition—imp

$$\langle \tilde{\mathbf{f}}_i, \mathbf{f}_{i'} \rangle = \delta[i - i'] \ . \tag{7.65}$$

This is seen as follows. There are four cases to consider. Assume for example that both i and i' are even. Then we have, using a change of variables,

$$\langle \tilde{\mathbf{f}}_{2m}, \mathbf{f}_{2n} \rangle = \sum_k g_0[k - 2m]h_0[2n - k]$$

$$= \sum_{k'} g_0[k']h_0[2(n - m) - k']$$

$$= \delta[n - m] \ .$$

The remaining three cases are obtained by similar computations, see Exer. 7.6.

One may ask whether the stability condition (7.49) also follows from the perfect reconstruction property. Unfortunately, this is not the case. This is one of the deeper mathematical results that we cannot cover in this book. A substantial background in Fourier analysis and functional analysis is needed to properly understand this question. We refer to the books [4, 5] for readers with the necessary mathematical background.

Now we will use the above results to define an important class of filters. We say that a set of four filters with the perfect reconstruction property are *orthogonal filters*, if the associated basis vectors constitute an orthonormal basis. By the above results and definitions this means that we must have

$\mathbf{f}_n = \tilde{\mathbf{f}}_n$ for all $n \in \mathbf{Z}$, or, translating this condition back to the filters using (7.63) and (7.64), that

$$g_0[k] = h_0[-k], \quad g_1[k] = h_1[-k], \tag{7.66}$$

for all $k \in \mathbf{Z}$.

Finally, let us note that if we start with a biorthogonal basis pair with the special structure imposed in (7.63) and (7.64), or with four filters satisfying the four equations (7.56), (7.58), (7.60), and (7.62), then we can reverse the arguments and show that the corresponding z-transforms $H_0(z)$, $H_1(z)$, $G_0(z)$, and $G_1(z)$ satisfy (7.31) and (7.32). The details are left as Exer. 7.7.

7.6 Summary of Results on Lifting and Filters

We now briefly summarize the above results. We have four different approaches to a building block for the DWT. They are

1. The building block is defined using lifting steps, and a final normalization step.
2. The building block is based on a 2×2 matrix (7.24), whose entries are Laurent polynomials, and on the scheme described in Fig. 7.3. The matrix is assumed to be invertible. This is the polyphase representation.
3. The building block is based on a two channel filter bank with the perfect reconstruction property, as in Fig. 7.4. The perfect reconstruction property can be specified in two different manners.
   a) The conditions (7.31) and (7.32) are imposed on the z-transforms of the filters, together with the determinant condition (7.37).
   b) The four filters satisfy the four equations (7.56), (7.58), (7.60), and (7.62) in the time domain.
4. Four filters are specified in the time domain, the vectors $\mathbf{f}_n$, $\tilde{\mathbf{f}}_n$, are defined as in (7.63) and (7.64), and it is required that they satisfy the biorthogonality condition.

The results in the previous sections show that all four approaches are equivalent, and we have also established formulas for translating between the different approaches. Part of the details were postponed to Chap. 12, namely the only not-so-easy translation: From filters to lifting steps.

7.7 Properties of Orthogonal Filters

We will now state a result on properties of orthogonal filters. Let us note that the orthogonality condition (7.66) and the equations (7.35) and (7.36) imply that we can only specify one filter. The remaining three filters are then determined by these conditions, up to the constant and odd time shift in

(7.35) and (7.36). We assume that the given filter is $\mathbf{h}_0$. To avoid the trivial case we also assume that its length L is at least 2.

Proposition 7.7.1. *Let $\mathbf{h}_0$, $\mathbf{h}_1$, $\mathbf{g}_0$, and $\mathbf{g}_1$ be four FIR filters defining a two channel filter bank with the perfect reconstruction property. Assume that the filters are orthogonal, i.e. they satisfy (7.66), and that they have real coefficients. Then the following results hold.*

1. *The length of $\mathbf{h}_0$ is even, $L = 2K$.*
2. *We can specify H_1 by*

$$H_1(z) = -z^{-2K-1}H_0(-z^{-1}) . \qquad (7.67)$$

If the nonzero entries in $\mathbf{h}_0$ are $h_0[1], \ldots, h_0[2K]$, then (7.67) shows that the nonzero entries in $\mathbf{h}_1$ are given by

$$h_1[k] = (-1)^k h[2K + 1 - k], \quad k = 1, 2, \ldots, 2K . \qquad (7.68)$$

3. *The power complementarity equation*

$$|H_0(e^{j\omega})|^2 + |H_1(e^{j\omega})|^2 = 2 \qquad (7.69)$$

holds for all $\omega \in \mathbf{R}$.
4. *$H_0(-1) = 0$ implies $H_0(1) = \sqrt{2}$.*
5. *Energy is conserved in the transition from $X(z)$ to $Y_0(z)$, $Y_1(z)$, see Fig. 7.4. In the time domain this is the equation*

$$\|\mathbf{x}\|^2 = \|\mathbf{y}_0\|^2 + \|\mathbf{y}_1\|^2 . \qquad (7.70)$$

We will now establish these five properties. Using the orthogonality condition (7.66) and the result (7.56), we find after changing the summation variable to $-k$ the result

$$\sum_k h_0[k]h_0[2n + k] = \delta[n] \quad \text{for all } n \in \mathbf{Z} . \qquad (7.71)$$

Assume now that the length of $\mathbf{h}_0$ is odd, $L = 2N + 1$. We can assume that the filter has the form

$$\mathbf{h}_0 = [\ldots, 0, h_0[0], \ldots, h_0[2N], 0, \ldots]$$

with $h_0[0] \neq 0$ and $h_0[2N] \neq 0$. Using (7.71) with $n = N \neq 0$ (since we have excluded the trivial case $L = 1$) we find $h_0[0]h_0[2N] = 0$, which is a contradiction.

Concerning the second result, then the assumption $g_0[k] = h_0[-k]$ implies $G_0(z) = H_0(z^{-1})$. Using (7.35) for $-z$ we then find

$$H_1(z) = -c^{-1}z^{-2k-1}G_0(-z) = -c^{-1}z^{-2k-1}H_0(-z^{-1}) .$$

We can choose $c = 1$ and $k = K$. Then (7.67) holds. Translating this formula to the time domain yields (7.68).

To prove the third result we go back to (7.55) and insert $G_0(z) = H_0(z^{-1})$ to get

$$H_0(z)H_0(z^{-1}) + H_0(-z)H_0(-z^{-1}) = 2 \ . \tag{7.72}$$

If we now take $z = e^{j\omega}$, and use that all coefficients are real, we see that the third result follows from this equation and (7.67).

Taking $z = 1$ in (7.72) yields the fourth result, up to a global choice of signs. We have chosen the plus sign.

The energy conservation property in (7.70) is much more complicated to establish. We recommend the reader to skip the remainder of this section on a first reading.

We use the orthogonality assumption to derive the equation

$$H_0(z)H_1(z^{-1}) + H_0(-z)H_1(-z^{-1}) = 0 \tag{7.73}$$

from (7.59). If we combine (7.67) and (7.72), we also find

$$H_1(z)H_1(z^{-1}) + H_1(-z)H_1(-z^{-1}) = 2 \ . \tag{7.74}$$

Taking $z = e^{j\omega}$ and using that the complex conjugate then is z^{-1}, we get from (7.72), (7.73), and (7.74) that for each $\omega \in \mathbf{R}$ the matrix

$$\mathbf{U}(e^{j\omega}) = \frac{1}{\sqrt{2}} \begin{bmatrix} H_0(e^{j\omega}) & H_0(e^{j(\omega+\pi)}) \\ H_1(e^{j\omega}) & H_1(e^{j(\omega+\pi)}) \end{bmatrix}$$

is a unitary matrix, since the three equations show that for a fixed $\omega \in \mathbf{R}$ the two rows have norm 1 (we include the $1/\sqrt{2}$ factor for this purpose) and are orthogonal, as vectors in $\mathbf{C}^2$.

The next step is to note that this matrix allows us to write the equations (7.28) and (7.29) as follows, again with $z = e^{j\omega}$,

$$\begin{bmatrix} Y_0(e^{j\omega}) \\ Y_1(e^{j\omega}) \end{bmatrix} = \mathbf{U}(e^{j\omega/2}) \begin{bmatrix} \frac{1}{\sqrt{2}} X(e^{j\omega/2}) \\ \frac{1}{\sqrt{2}} X(e^{j(\omega/2+\pi)}) \end{bmatrix} .$$

Note how we have distributed the factor $1/2$ in (7.28) and (7.29) over the two terms. The unitarity of $\mathbf{U}(e^{j\omega/2})$ for each ω means that this transformation preserves the norm in $\mathbf{C}^2$. Thus we have

$$|Y_0(e^{j\omega})|^2 + |Y_1(e^{j\omega})|^2 = \frac{1}{2} \left(|X(e^{j\omega/2})|^2 + |X(e^{j(\omega/2+\pi)})|^2 \right) \tag{7.75}$$

for all $\omega \in \mathbf{R}$.

The last step consists in using Parseval's equation (7.2), then (7.75), and a change in integration variable from $\omega/2$ to ω, and finally Parseval's equation once more.

$$\|\mathbf{y}_0\|^2 + \|\mathbf{y}_1\|^2 = \frac{1}{2\pi} \int_0^{2\pi} \left(|Y_0(e^{j\omega})|^2 + |Y_1(e^{j\omega})|^2 \right) d\omega$$

$$= \frac{1}{4\pi} \int_0^{2\pi} \left(|X(e^{j\omega/2})|^2 + |X(e^{j(\omega/2+\pi)})|^2 \right) d\omega$$

$$= \frac{1}{2\pi} \int_0^{\pi} \left(|X(e^{j\omega})|^2 + |X(e^{j(\omega+\pi)})|^2 \right) d\omega$$

$$= \frac{1}{2\pi} \int_0^{2\pi} |X(e^{j\omega})|^2 d\omega$$

$$= \|\mathbf{x}\|^2 .$$

This computation concludes the proof of the proposition.

7.8 Some Examples

Let us now apply the above results to some of the previously considered transforms. We start as usual with the Haar transform. It is given by (3.28)–(3.31). In the notation introduced here (compare with the example in Sect. 7.2) we get

$$\mathbf{H}(z) = \begin{bmatrix} \sqrt{2} & 0 \\ 0 & \frac{1}{\sqrt{2}} \end{bmatrix} \begin{bmatrix} 1 & \frac{1}{2} \\ 0 & 1 \end{bmatrix} \begin{bmatrix} 1 & 0 \\ -1 & 1 \end{bmatrix} .$$

Multiplying these matrices we get

$$\mathbf{H}(z) = \begin{bmatrix} \frac{1}{\sqrt{2}} & \frac{1}{\sqrt{2}} \\ -\frac{1}{\sqrt{2}} & \frac{1}{\sqrt{2}} \end{bmatrix} .$$

Using (7.38) and (7.39) we get

$$H_0(z) = \frac{1}{\sqrt{2}}(1+z) ,$$

$$H_1(z) = \frac{1}{\sqrt{2}}(-1+z) ,$$

and then in the frequency variable (on the unit circle in the complex plane)

$$|H_0(e^{j\omega})| = \sqrt{2}|\cos(\omega/2)| ,$$
$$|H_1(e^{j\omega})| = \sqrt{2}|\sin(\omega/2)| .$$

The graphs of these two functions have been plotted in Fig. 7.5. We see that H_0 is a low pass filter, but not very sharp. Note that we have chosen a linear scale on the vertical axis. Often a logarithmic scale is chosen in this type

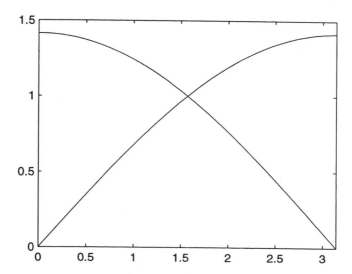

Fig. 7.5. Plot of $|H_0(e^{j\omega})|$ and $|H_1(e^{j\omega})|$ for the Haar transform

of plots. We will now make the same computations for the Daubechies 4 transform. It is given by lifting steps in (3.18)–(3.22). Written in matrix notation we have

$$\mathbf{H}(z) = \begin{bmatrix} \frac{\sqrt{3}-1}{\sqrt{2}} & 0 \\ 0 & \frac{\sqrt{3}+1}{\sqrt{2}} \end{bmatrix} \begin{bmatrix} 1 & -z \\ 0 & 1 \end{bmatrix} \begin{bmatrix} 1 & 0 \\ -\frac{\sqrt{3}}{4} - \frac{\sqrt{3}-2}{4}z^{-1} & 1 \end{bmatrix} \begin{bmatrix} 1 & \sqrt{3} \\ 0 & 1 \end{bmatrix}.$$

Multiplying we find

$$\mathbf{H}(z) = \begin{bmatrix} \frac{3-\sqrt{3}}{4\sqrt{2}}z + \frac{1+\sqrt{3}}{4\sqrt{2}} & -\frac{\sqrt{3}-1}{4\sqrt{2}}z + \frac{3+\sqrt{3}}{4\sqrt{2}} \\ -\frac{3+\sqrt{3}}{4\sqrt{2}} + \frac{\sqrt{3}-1}{4\sqrt{2}}z^{-1} & \frac{\sqrt{3}+1}{4\sqrt{2}} + \frac{3-\sqrt{3}}{4\sqrt{2}}z^{-1} \end{bmatrix}.$$

Using the equations (7.38) and (7.39) we get the Daubechies 4 filters

$$H_0(z) = h[0] + h[1]z + h[2]z^2 + h[3]z^3 \ ,$$
$$H_1(z) = -h[3]z^{-2} + h[2]z^{-1} - h[1] + h[0]z \ ,$$

where

$$h[0] = \frac{1+\sqrt{3}}{4\sqrt{2}} \ , \qquad h[1] = \frac{3+\sqrt{3}}{4\sqrt{2}} \ ,$$

$$h[2] = \frac{3-\sqrt{3}}{4\sqrt{2}} \ , \qquad h[3] = \frac{1-\sqrt{3}}{4\sqrt{2}} \ . \tag{7.76}$$

Note that while obtaining the filter taps from the lifting steps is easy, since it is just a matter of matrix multiplications, it is by no means trivial to go the other way. This is described in details in Chap. 12.

The absolute values of $H_0(e^{j\omega})$ and $H_1(e^{j\omega})$ have been plotted in Fig. 7.6. We note that these filters give a sharper cutoff than the Haar filters.

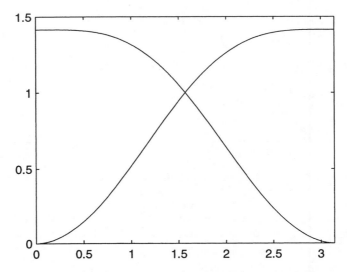

Fig. 7.6. Plot of $|H_0(e^{j\omega})|$ and $|H_1(e^{j\omega})|$ for the Daubechies 4 transform

Finally we repeat these computations for the CDF(2,2) transform. It is given in the lifting form by equations (3.13) and (3.14) and the normalization by (3.40) and (3.41). We then get

$$\mathbf{H}(z) = \begin{bmatrix} \sqrt{2} & 0 \\ 0 & \frac{1}{\sqrt{2}} \end{bmatrix} \begin{bmatrix} 1 & \frac{1}{4}(1 + z^{-1}) \\ 0 & 1 \end{bmatrix} \begin{bmatrix} 1 & 0 \\ -\frac{1}{2}(1 + z) & 1 \end{bmatrix}.$$

Multiplying these terms yields

$$\mathbf{H}(z) = \begin{bmatrix} \frac{3}{2\sqrt{2}} - \frac{1}{4\sqrt{2}}z - \frac{1}{4\sqrt{2}}z^{-1} & \frac{1}{2\sqrt{2}} + \frac{1}{2\sqrt{2}}z^{-1} \\ -\frac{1}{2\sqrt{2}} - \frac{1}{2\sqrt{2}}z & \frac{1}{\sqrt{2}} \end{bmatrix}.$$

As above we compute

$$H_0(z) = -\frac{1}{4\sqrt{2}}z^2 + \frac{1}{2\sqrt{2}}z + \frac{3}{2\sqrt{2}} + \frac{1}{2\sqrt{2}}z^{-1} - \frac{1}{4\sqrt{2}}z^{-2},$$

$$H_1(z) = -\frac{1}{2\sqrt{2}}z^2 + \frac{1}{\sqrt{2}}z - \frac{1}{2\sqrt{2}}.$$

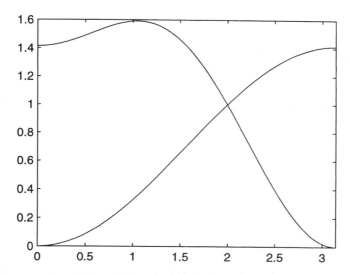

Fig. 7.7. Plot of $|H_0(e^{j\omega})|$ and $|H_1(e^{j\omega})|$ for the CDF(2,2) transform

The graphs are plotted in Fig. 7.7. The unusual frequency response is due to the fact that these filters are not orthogonal.

Let us finally comment on available filters. There is a family of filters constructed by I. Daubechies, which we refer to as the Daubechies filters and also the Daubechies transform, when we think of the associated DWT. The Haar filters and the Daubechies 4 transform discussed above are the first two members of the family. The filters are often indexed by their length, which is an even integer. All filters in the family are orthogonal. As the length increases they (slowly) converge to ideal low pass and high pass filters.

Another family consists of the so-called symlets. They were also first constructed by I. Daubechies. These filters are orthogonal and are also indexed by their length. They are less asymmetrical than the first family, with regard to phase response.

A third family of orthogonal filters with nice properties is the Coiflets. They are described in [5], where one also finds tables of filter coefficients.

We have already encountered some families of biorthogonal filters. They come from the CDF-families described in Sect. 3.6, and in the notation CDF(N,M) the integers N and M denote the multiplicity of the zero at $z = -1$ for the z-transforms $H_0(z)$ and $G_0(z)$, respectively. We say that $H_0(z)$ has a zero of multiplicity N at $z = -1$, if there exists a Laurent polynomial $\tilde{H}_0(z)$ with $\tilde{H}_0(-1) \neq 0$, such that $H_0(z) = (z+1)^N \tilde{H}_0(z)$.

The various toolboxes have functions for generating filter coefficients. We will discuss those in *Uvi_Wave* in some detail later.

Exercises

7.1 Carry out explicitly the change of summation variable, which proves (7.9) in the time domain.

7.2 Verify that the update step in the CDF(2,2) transform example in Sect. 7.2 is given by $S(z) = \frac{1}{4}(1 + z^{-1})$.

7.3 Let $\mathbf{h}$ be a filter. Show that filtering by $\mathbf{h}$ preserves the finite energy property, i.e. $\|\mathbf{h} * \mathbf{x}\| \leq c\|\mathbf{x}\|$ for all $\mathbf{x} \in \ell^2(\mathbf{Z})$, by using the z-transform and Parseval's equation.

7.4 Verify that in the polyphase representation from Sect. 7.2 the invertibility of the matrix (7.24) is equivalent with the perfect reconstruction property in the two channel filter bank case.

7.5 Verify that the canonical basis defined in (7.48) satisfies all the requirements for an orthonormal basis for $\ell^2(\mathbf{Z})$.

7.6 Carry out the remaining three cases in the verification of (7.65).

7.7 Let four filters $\mathbf{h}_0$, $\mathbf{h}_1$, $\mathbf{g}_0$, and $\mathbf{g}_1$ satisfy the equations (7.56), (7.58), (7.60), and (7.62). Show that their z-transforms satisfy (7.31) and (7.32).

7.8 Go through the details in establishing the formulas (7.51), (7.52), and (7.53). Try also to obtain these formulas from the polyphase formulation of the filter bank.

7.9 Verify that (7.67) leads to (7.68).

7.10 Show that (7.53) is valid if and only if

$$\sum_n \left(g_0[k - 2n]h_0[2n - m] + g_1[k - 2n]h_1[2n - m] \right) = \delta[m - k] \,,$$

and then show that this is true for all wavelet filters.

7.11 Carry out computations similar to those in Sect. 7.8 for the CDF(3,1) transform defined in Sect. 3.6.

8. Wavelet Packets

In the first chapters we have introduced the lifting technique, which is a method for defining and implementing the discrete wavelet transform. The definitions were given in Chap. 3, and simple examples were given in Chap. 4. In particular, we saw applications of the wavelet analysis to noise reduction. The applications were based on the one scale building block, applied several times. In this chapter we want to extend the use of these building blocks to define many new transforms, all called *wavelet packets*. In many applications these new transforms are the basis for the successful use of wavelets. Some examples are given in Chap. 13.

8.1 From Wavelets to Wavelet Packets

In this chapter we regard the one scale DWT as a given building block and we extend the concept of a wavelet analysis. Thus we regard the one scale DWT as a black box, which can act on any signal of even length. The analysis transform is now denoted by $\mathbf{T}_a$ and the inverse synthesis transform by $\mathbf{T}_s$.

We recall that the direct transform is capable of transforming *any* signal of even length into two signals, and the inverse transform reverses this decomposition, i.e. we are requiring the perfect reconstruction property of our building block.

Let the signal be represented by a box, with a length proportional to the length of the signal. A transformation followed by an inverse transformation will then be represented as in Fig. 8.1.

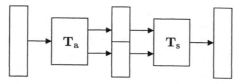

Fig. 8.1. Block diagram for building block and its inverse, with signal elements shown

Three consecutive transforms will look like Fig. 8.2(a). The dotted boxes show the signal parts transferred without transformation from one application of the building block to the next. Other examples are in Fig. 3.7, and in Table 2.1 and 3.1, which show the same with numbers and symbols. Note that the orientation and location of the transformed components are different here and in the examples mentioned.

Any collection of consecutive transforms of a given signal is called a *decomposition* of this signal. We use the following terminology. The original signal is said to be at the first level. Applying the transform once gets us to the second level. Thus in the wavelet transform case the k scale transform leads to a decomposition with $k + 1$ levels.

Note that the original signal always is the top level in a decomposition. Depending on how we draw the diagrams, the top level is on the left, or on the top of the diagram. Note that the placement of the **s** and **d** parts of each transform step agrees with the diagram in Fig. 3.7.

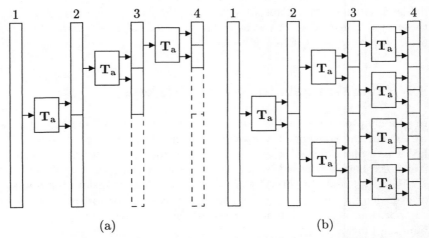

Fig. 8.2. The wavelet (a) and the full wavelet packet (b) decompositions of a signal

When looking at Fig. 8.2(a), one starts to wonder, why some signal parts are transformed and other parts are not. One can apply the transform to all parts, or, as we shall see soon, to selected parts. This idea is called *wavelet packets*, and a full wavelet packet decomposition over four levels is shown in Fig. 8.2(b). Note that in this case we have applied the transform $\mathbf{T_a}$ 7 times. The standard abbreviation WP will be used in the rest of the text. Each signal part box in Fig. 8.2 is called an *element*. Note that this word might also be used, when it is clear from the context, to denote a single number or coefficient in a signal.

In Chap. 4 we saw the wavelet decomposition applied to problems of noise reduction, and in the separation of slow and fast variations in a signal. But, by increasing the number of possible decompositions, we may be able to do better, in particular with signals, where the usual wavelet decomposition does not perform well.

Let us go back to the very first example in Sect. 2.1. We saw that a signal containing a total of 15 digits could be transformed into one containing 13 digits, and then in the next two steps to signals with 13 and 12 digits, respectively. We can get from one of these four representations to any other, by applying the direct or inverse transforms the right number of times. They are equivalent representations of the original signal. So if we want to reduce memory usage, we choose the one requiring the least amount of memory space. Now if we use the wavelet packet decompositions, then we have 26 different, but equivalent, representations to choose from. Thus the chances of getting an efficient representation will be greater.

Though the number 26 seems rather unmotivated, it has been carefully deduced. But before elaborating on this we will see how a representation can be extracted from the WP decomposition. As an example we use the same signal as in the first chapters. The full WP decomposition is given in Table 8.1. A representation is a choice of elements, sequentially concatenated,

Table 8.1. Full wavelet packet decomposition of the signal

56	40	8	24	48	48	40	16
48	16	48	28	8	−8	0	12
32	38	16	10	0	6	8	−6
35	−3	13	3	3	−3	1	7

such that

1. the selected elements cover the original signal, and
2. there is no overlap between the selected elements.

The first condition ensures sufficient information for reconstruction, while the second one ensures that no unnecessary information is chosen. Both conditions are needed. Since any representation is equivalent with a change of basis (here we view the original signal as given in the canonical basis), a choice of representation corresponds to a choice of basis (this is elaborated in Chap. 5). We will, in agreement with most of the wavelet literature, often use the formulation 'choice of basis.' But one should remember that a representation is different from a basis.

An example of a representation (notice how the two conditions are fulfilled) of the signal decomposed above is

48	16	48	28	3	−3	8	−6

.

The original signal is reconstructed by first using $\mathbf{T_s}$ on

| 3 | −3 | which becomes | 0 | 6 | .

Then $\mathbf{T_s}$ is used again on

| 0 | 6 | 8 | −6 | which becomes | 8 | −8 | 0 | 12 | .

Finally the original signal is recreated by one more $\mathbf{T_s}$. Graphically it looks like Fig. 8.3, where the $\mathbf{T_a}$ and $\mathbf{T_s}$-boxes have been left out.

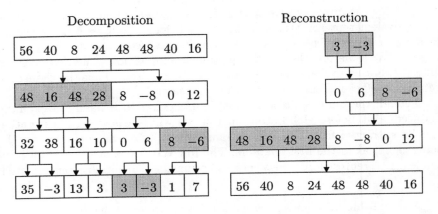

Fig. 8.3. Decomposition of the signal, and reconstruction from one particular representation.

To further exemplify the decomposition process, six other representations are shown in Fig. 8.4. In Sect. 11.7 it is demonstrated how a WP decomposition can be implemented in MATLAB.

8.2 Choice of Basis

With the extension of wavelets to wavelet packets the number of different representations increases significantly. We claimed for example that the full decomposition of the signal given in the first section gave a total of 26 possible representations. As we shall see below, this number grows very rapidly, when more levels are added to the decomposition.

We now assume that we have decided on a criterion for choosing the best basis, a so-called cost function. It could be the number of digits used in the representation of the signal. With 26 representations we can inspect each one to find the best representation, but this becomes an overwhelming task with one billion representations (which is possible with 7 levels instead of 4). It is therefore imperative to find a method to assist us in choosing the best basis. The method should preferably be both fast and exhaustive, such the

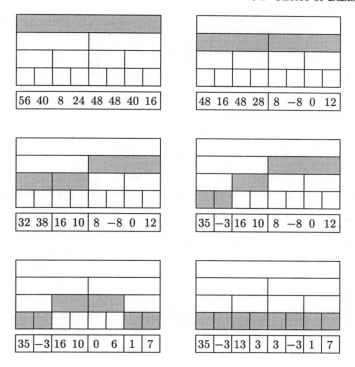

Fig. 8.4. Six different representation of the signal.

chosen basis is the best one, according to our criterion. If more than one basis satisfies our criterion for best basis, the method should find one of them. We present such an algorithm below, in Sect. 8.2.2.

8.2.1 Number of Bases

It is not very difficult to set up an equation for the number of basis in a decomposition with a given number of levels. We start with a decomposition containing $j+1$ levels. The number of bases in this decomposition is denoted by A_j. This makes the previous claim equivalent to $A_3 = 26$. Now, a decomposition with $j+2$ levels has A_{j+1} bases, and this number can be related to A_j by the following observation. The larger decomposition can be viewed as consisting of three parts. Two of them are smaller decompositions with $j+1$ levels each. The third part is the original signal. See Fig. 8.5. Every time we choose a basis in the left of the two small decomposition, there are A_j choices in the right, since we can combine left and right choices freely. A total of A_j^2 different bases can be chosen this way. For any choice of elements in the two smaller decompositions, the original signal cannot be chosen, since no overlap is allowed. If we do choose the top element, however, this also

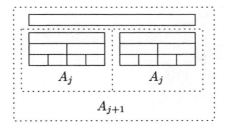

Fig. 8.5. Counting the number of decompositions

counts as a representation. Thus a decomposition with $j + 2$ levels satisfies $A_{j+1} = A_j^2 + 1$. Starting with $j = 0$, we find this to be a decomposition with one level, so $A_0 = 1$. Then $A_1 = 1^2 + 1 = 2$, and $A_2 = 2^2 + 1 = 5$, and then $A_3 = 5^2 + 1 = 26$, the previous claim.

Continuing a little further, we are in for a surprise, see Table 8.2.

Table 8.2. Growth in the number of decompositions with level

Number of levels	Minimum signal length	Number of bases
1	1	1
2	2	2
3	4	5
4	8	26
5	16	677
6	32	458330
7	64	210066388901
8	128	44127887745906175987802

The number of bases grows extremely fast, approximately doubling the number of digits needed to represent it, for each extra level. It is worth noting that the number of bases does not depend directly on the length of the signal. A signal of length 128 transformed 3 times, to make a decomposition with 4 levels, has only 26 different representations. The number of levels, we can consider, is limited by the length of the signal, since the decomposition must terminate, when an element contains just one number. The minimum length required for a given level is shown in the second column of Table 8.2, in the first few cases. In general, decomposition into J levels requires a signal of length at least 2^{J-1}.

We can find an upper and a lower bound for A_j. Take two other equations

$$B_{j+1} = B_j^2, \quad B_1 = 2 \quad \text{and} \quad C_{j+1} = C_j^4, \quad C_1 = 2.$$

Clearly $B_j \le A_j \le C_j$ for $j > 0$. Both B_j and C_j are easily found. For B_j we have

$$B_j = B_{j-1}^2 = \left(B_{j-2}^2\right)^2$$

$$= \underbrace{\left(\left(B_1^2\right)^2 \cdots\right)^2}_{j-1 \text{ powers of } 2} = \left(\left(2^2\right)^2 \cdots\right)^2 = 2^{2 \cdot 2 \cdots \cdot 2} = 2^{2^{j-1}}.$$

An analogous computation shows that $C_j = 2^{2^j}$. Hence we have the bounds

$$2^{2^{j-1}} < A_j < 2^{2^j}.$$

For a decomposition with 10 levels, i.e. $j = 9$, we have a lower bound of $2^{2^8} \approx 10^{77}$ and an upper bound of $2^{2^9} \approx 10^{154}$. These numbers are very large.

8.2.2 Best Basis

The concept of the best basis depends on the application we have in mind. To use the concept in an algorithm, we introduce a cost function. A cost function measures in some terms the cost of a given representation, with the idea that the best basis is the one having the smallest cost.

To be usable in an algorithm, the cost function must have some specific properties. We denote a cost function by the symbol $\mathcal{K}$ here. The cost function is defined on finite vectors of arbitrary length. The value of the cost function is a real number. Given two vectors of finite length, $\mathbf{a}$ and $\mathbf{b}$, we denote their concatenation by $[\mathbf{a}\,\mathbf{b}]$. This vector simply consists of the elements in $\mathbf{a}$ followed by the elements in $\mathbf{b}$. We require the following two properties.

1. The cost function is additive in the sense that $\mathcal{K}([\mathbf{a}\,\mathbf{b}]) = \mathcal{K}(\mathbf{a}) + \mathcal{K}(\mathbf{b})$ for all finite length vectors $\mathbf{a}$ and $\mathbf{b}$.
2. $\mathcal{K}(\mathbf{0}) = 0$, where $\mathbf{0}$ denotes the zero vector.

As an example, we take the cost function, which counts the number of nonzero entries in a vector. For example,

$$\mathcal{K}([1\ 5\ 0\ -3\ 4\ 0\ 0\ -6]) = 5.$$

The additivity is illustrated with this example

$$\mathcal{K}([1\ 5\ 0\ -3\ 4\ 0\ 0\ -6]) = \mathcal{K}([1\ 5\ 0\ -3]) + \mathcal{K}([4\ 0\ 0\ -6]).$$

The conditions on a cost function can be relaxed in some cases, where the structure of the signal is known, and a near-best basis is acceptable. We will not pursue this topic in this book.

Let us now describe the algorithm, which for a given cost function finds a best basis. The starting point is a computation of the full wavelet packet decomposition to a prescribed level J, compatible with the length of the signal. An example is shown in Fig. 8.2(b), with $J = 4$. The next step is to calculate the cost values of all elements of the full decomposition. Note that these two computations are performed only once. Their complexity is proportional to the length of the given signal multiplied by the number of levels chosen in the full decomposition.

Given this full decomposition with the cost of each element computed, the algorithm performs a bottom-up search in this tree. It can be described as follows.

1. Mark all elements on the bottom level J.
2. Let $j = J$.
3. Let $k = 0$.
4. Compare the cost value v_1 of element k on level $j - 1$ (counting from the left on that level) to the sum v_2 of the cost values of the elements $2k$ and $2k + 1$ on level j.
    a) If $v_1 \leq v_2$, all marks below element k on level $j - 1$ are deleted, and element k is marked.
    b) If $v_1 > v_2$, the cost value v_1 of element k is replaced with v_2.
5. $k = k + 1$. If there are more elements on level j (if $k < 2^{j-1} - 1$), go to step 4.
6. $j = j - 1$. If $j > 1$, go to step 3.
7. The marked basis has the lowest possible cost value, which is the value currently assigned to the top element.

The additivity ensures that the algorithm quickly finds a best basis, which of course need not be unique. Note that once the first two steps (full decomposition and computation of cost) have been performed, then the complexity of the remaining computations only depends on the number of levels J being used. The complexity of this part is found to be $O(J \log J)$.

The algorithm is most easily understood with the help of an example. For that purpose we reuse the decomposition given in Table 8.1. First the cost value of each element is calculated. The cost values are represented in the same tree structure as the full decomposition. As cost function in this example we choose the count of numbers with absolute value > 1. Calculated cost values and the marking of the bottom level are both shown in Fig. 8.7(1). We start with the bottom level, since the search starts here. We then move up each time it is possible to reduce total cost by doing so. The additivity makes partial replacement of elements possible. The remaining crucial step is the comparison of cost values. All comparisons are between an element and the two elements just below it. If the sum of the cost values in the two lower elements (v_2 in the algorithm) is smaller than the cost value in the upper element (v_1 in the algorithm), this sum is inserted as a new cost value in the upper element. This possibility is illustrated in the top row of Fig. 8.6. If, on the other hand, the sum is larger than (or equal to) the cost value in the upper element, this element is marked and all marks below this element are deleted. This possibility is illustrated in the bottom row of Fig. 8.6.

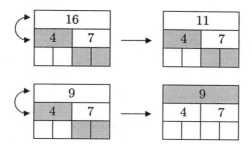

Fig. 8.6. Comparison of cost values in the two cases

All elements on each level is run through, and the levels are taken from the lowest but one and up. In Fig. 8.7 the process is shown in four steps. Notice

56	40	8	24	48	48	40	16
48	16	48	28	8	−8	0	12
32	38	16	10	0	6	8	−6
35	−3	13	3	3	−3	1	7

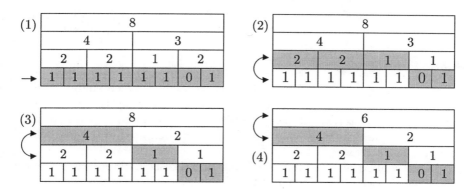

Fig. 8.7. An example showing the best basis search. The cost function is the count of numbers in each element with absolute value > 1

how the cost value in the top element is the cost value of the best basis at the end of the search. The best representation of this signal has been found to be

48	16	48	28	0	6	1	7

.

In the left half of the decomposition the cost value is 4 in all of the 5 possible choices of elements. This means that with respect to cost value 5 different best bases exist in this decomposition. The algorithm always finds one of these. The equal sign in step 4(a) has as a consequence that the best basis with fewest transform steps is picked. Below we will use the term 'best basis' for the one selected by this algorithm.

If we had chosen the cost function to be the count of numbers with absolute value > 3, the best basis would have been the eight elements on the bottom level, with a total cost value of only 3.

One particular type of best basis is the *best level basis*, where the chosen basis consists solely of elements from one level. The number of best level bases is equal to the number of levels. This type of basis is often used in time-frequency planes.

In Sect. 11.8 it is shown how to implement the best basis algorithm in MATLAB.

8.3 Cost Functions

Any function defined on all finite length vectors, and taking real values, can be used as a cost function, if it satisfies the two conditions on page 93. Some of the more useful cost functions are those which measure concentration in some sense. Typically the concentration is measured in norm or entropy. Low cost, and consequently high concentration, means in these cases that there are few large and many small numbers in an element. We start with a simple example of a cost function, which also can be thought of as measuring concentration.

8.3.1 Threshold

One of the simplest cost functions is the threshold, which simply returns the count of numbers above a specified threshold. Usually the sign of each number is of no interest, and the count is therefore of numbers with absolute value above the threshold. This was the cost function used in the example in the previous section.

In the context of a cost function given by a threshold, 'large' means above the given threshold, and 'small' below the threshold. Low cost then means that the basis represents the signal with as few large values as possible. In many cases the large values are the significant values. This is often the case, when one uses the wavelet transform.

But there are certain pitfalls in this argument. The following situation can easily arise. A signal with values in the range -1 to 1 is transformed into two signal each with values in the range -2 to 2. One more transform would make the range -4 to 4, and so on. It is apparent from the plots in Fig. 7.5, 7.6, and 7.7 that the one level wavelet transform has a gain above 1. But this does not mean that we gain information just by transforming the signal. The increase in the range of values is due to the normalization of the one level DWT, which in the orthogonal case means that we have energy preserved during each step in the decomposition.

Let us give an example. Take the vector $\mathbf{a} = [1\ 1\ 1\ 1\ 1\ 1\ 1\ 1]$. Then $\|\mathbf{a}\| = \sqrt{8}$. Let $\mathbf{b}$ denote its transform under an energy preserving transformation, such that $\|\mathbf{b}\| = \sqrt{8}$. Assume now that the transform has doubled the range of the signal. This means that at least one of the entries has to be ± 2. Thus one could find for example $\mathbf{b} = [2\ 2\ 0\ 0\ 0\ 0\ 0\ 0]$. Since energy is preserved, at most two samples can have absolute value equal to 2. So the overall effect is that most of the entries actually decrease in amplitude. This effect explains, at least partially, why the wavelet transform is so useful in signal compression, and in other applications.

The threshold is a very simple cost function. To satisfy the additivity property the threshold has to be the same for all levels. Furthermore, an inappropriate choice of threshold value can lead to a futile basis search. If the threshold is chosen too high, then all cost values will be zero, and the basis search returns the original signal. The same is the case, if the threshold is too low.

8.3.2 ℓ^p-Norm and Shannon's Entropy

The problem with the threshold cost function just mentioned leads one to look for cost functions with better properties. In this section we describe two possibilities. The first one is the so-called ℓ^p-norm. The second one is a modified version of Shannon's entropy. Both have turned out to be very useful in applications. The two cost functions are defined as follows.

Definition 8.3.1. *For* $0 < p < \infty$ *the cost function based on the* ℓ^p*-norm is given by*

$$\mathcal{K}_{\ell^p}(\mathbf{a}) = \sum_n |a[n]|^p \qquad (8.1)$$

for all vectors $\mathbf{a}$ *of finite length.*

We see that the energy in the signal is measured by the cost function $\mathcal{K}_{\ell^2}$. If we use this cost function together with a one scale DWT, which preserves energy, then we find that the best basis search algorithm always returns the original representation. But for $0 < p < 2$ this cost function, together with an energy preserving transform, can be very useful. The reason is that $\mathcal{K}_{\ell^2}(\mathbf{a}) = \mathcal{K}_{\ell^2}(\mathbf{b})$ and $\mathcal{K}_{\ell^p}(\mathbf{a}) < \mathcal{K}_{\ell^p}(\mathbf{b})$ together imply that the vector $\mathbf{a}$ must contain fewer large elements than $\mathbf{b}$. See Exer. 8.4.

Definition 8.3.2. *The cost function based on Shannon's entropy is defined by*

$$\mathcal{K}_{\text{Shannon}}(\mathbf{a}) = -\sum_n a[n]^2 \log(a[n]^2) \qquad (8.2)$$

for all vectors $\mathbf{a}$ *of finite length. Here we use the convention* $0\log(0) = 0$.

Let us note that this cost function is not the entropy function, but a modified one. The original entropy function is for a signal **a** computed as

$$-\sum_n p[n]\log(p[n]), \quad \text{where } p[n] = a[n]^2/\|\mathbf{a}\|^2 \ .$$

This function fails to satisfy the additivity condition due to the division by $\|\mathbf{a}\|$. But the cost function defined here has the property that its value is minimized, if and only the original entropy function is minimized.

Let us show that the cost function $\mathcal{K}_{\text{Shannon}}$ measures concentration in a signal by an example. Let $\mathbf{1}_N$ denote the vector of length N, all of whose entries equal 1. Let $\mathbf{s}_N = (E/N)^{1/2}\mathbf{1}_N$, i.e. all entries equal $(E/N)^{1/2}$. The energy in this signal is equal to E, independent of N, while $\mathcal{K}_{\text{Shannon}}(\mathbf{s}_N) = -E\log(E/N)$. For a fixed E this function is essentially $\log(N)$. This shows that the entropy increases, if we distribute the energy in the signal evenly over an increasing number of entries. More generally, one can prove that for signals with fixed energy, the entropy attains a maximum, when all entries are equal, and a minimum, when all but one entry equal zero.

In Sect. 11.9 the implementation of the different cost functions is presented, and in Chap. 13 some examples of applications are given.

Exercises

8.1 Verify that the threshold cost function satisfies the two requirements for a cost function.

8.2 Verify that the cost function $\mathcal{K}_{\ell^p}$ satisfies the two requirements for a cost function.

8.3 Verify that the cost function $\mathcal{K}_{\text{Shannon}}$ satisfies the two requirements for a cost function.

8.4 Let $\mathbf{a} = [a[0]\ a[1]]$ and $\mathbf{b} = [b[0]\ b[1]]$ be two nonzero vectors of length 2 with nonnegative entries. Assume that $\mathcal{K}_{\ell^2}(\mathbf{a}) = \mathcal{K}_{\ell^2}(\mathbf{b})$, but $\mathcal{K}_{\ell^1}(\mathbf{a}) < \mathcal{K}_{\ell^1}(\mathbf{b})$. Assume that $b[0] = b[1]$. Show that either $a[0] < a[1]$ or $a[0] > a[1]$.

8.5 Assume that a full wavelet packet decomposition has been computed for a given signal, using an energy preserving transform. Take as the cost function $\mathcal{K}_{\ell^2}$. Go through the steps in the best basis search algorithm to verify that the algorithm selects the original signal as the best basis.

8.6 Assume that a full wavelet packet decomposition has been computed for a given signal. Assume that a threshold τ is chosen, which is larger that the largest absolute value of all elements in the decomposition. Choose the threshold cost function with this threshold. Go through the steps in the best basis search algorithm to verify that the algorithm selects the original signal as the best basis.

9. The Time-Frequency Plane

Time-frequency analysis is an important tool in modern signal analysis. By using information on the distribution of the energy in a signal with respect to both time and frequency, one hopes to gain additional insight into the nature of signals.

In this chapter we introduce the time-frequency plane as a tool for visualizing signals and their transforms. We look at the discrete wavelet transform, wavelet packet based transforms, and also at the short time Fourier transform. The connection between time and frequency is given by the Fourier transform, which we introduced in Chap. 7. We will need further results from Fourier analysis. They will be presented here briefly. What is needed can be found in the standard texts on signal analysis, or in many mathematics textbooks.

9.1 Sampling and Frequency Contents

We consider a discrete signal with finite energy, $\mathbf{x} \in \ell^2(\mathbf{Z})$. The frequency contents of this signal is given by the Fourier transform in the form of the associated Fourier series

$$X(\omega) = \sum_n x[n]e^{-jn\omega} \ . \tag{9.1}$$

See Chap. 7 for some results on Fourier series. The function $X(\omega)$ is periodic with period 2π, which means that $X(\omega+2\pi k) = X(\omega)$ for all $k \in \mathbf{Z}$. Therefore the function is completely determined by its values on an interval of length 2π. In this book we always take our signals to have real values. For a real signal $\mathbf{x}$ we have $\overline{X}(\omega) = X(-\omega)$, as can be seen by taking the complex conjugate of both sides in (9.1). As a consequence, the frequency contents is determined by the values of $X(\omega)$ on any interval of the form $[k\pi, (k+1)\pi]$, where k can be any integer. Usually one chooses the interval $[0, \pi]$.

To interpret our signals we need to fix units. The discrete signal is indexed by the integers. If we choose a time unit T, which we will measure in seconds, then we can interpret the signal as one being measured at times nT, $n \in \mathbf{Z}$. Let us assume that there is an underlying analog, or continuous, signal, such that the discrete signal has been obtained by sampling this continuous signal at times nT. The number $1/T$ is called the sampling rate, and $f_s = 2\pi/T$ the sampling frequency. Note that some textbooks use Fourier series based on the functions $\exp(-j2\pi n\omega)$. In those books the sampling frequency is often defined to be $1/T$.

If we now introduce the time unit explicitly in the Fourier series, then it becomes

$$X_T(\omega) = \sum_n x[n]e^{-jnT\omega} . \tag{9.2}$$

The function $X_T(\omega)$ is periodic with period $2\pi/T$. After a change of variables, Parseval's equation (7.2) reads

$$\sum_n |x[n]|^2 = \frac{T}{2\pi} \int_{-\pi/T}^{\pi/T} |X_T(\omega)|^2 \, d\omega . \tag{9.3}$$

For a real discrete signal the frequency contents is then determined by the values of $X_T(\omega)$ on for example the interval $[0, \pi/T]$. This is often expressed by saying that in a sampled signal one can only find frequencies up to half the sampling frequency, $\frac{1}{2}f_s$, which is equal to π/T by our definition. This result is part of Shannon's sampling theorem. See for example [5, 16, 22, 23, 28] for a discussion of this theorem.

As mentioned above, for a real signal we can choose other intervals in frequency, on which the values of $X_T(\omega)$ will determine the signal. Any interval of the form $[k\pi/T, (k+1)\pi/T]$ can be chosen. This is not a violation of the sampling theorem, but simply a consequence of periodicity, and the assumption that the signal is real. We have illustrated the possibilities in Fig. 9.1. The usual choice is marked by the heavy line segment. Other possibilities are marked by the thin line segments. Note how the symmetry $|X_T(\omega)| = |X_T(-\omega)|$ is also shown in the figure.

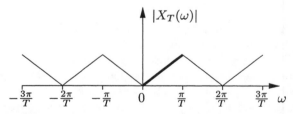

Fig. 9.1. Choice of frequency interval for a sampled signal. Heavy line marks usual choice

The frequency contents of an analog signal $x(t)$ is given by the continuous Fourier transform, which is defined as

$$\hat{x}(\omega) = \int_{-\infty}^{\infty} x(t)e^{-j\omega t}\,dt \,. \tag{9.4}$$

The inversion formula is

$$x(t) = \frac{1}{2\pi} \int_{-\infty}^{\infty} \hat{x}(\omega)e^{j\omega t}\,d\omega \,. \tag{9.5}$$

For a real signal we have $\overline{\hat{x}}(\omega) = \hat{x}(-\omega)$, such that it suffices to consider positive frequencies. Any positive frequency may occur. Suppose now we sample an analog signal at a rate $1/T$. This means that we take $x[n] = x(nT)$. Recall that we use square brackets for discrete variables and round brackets for continuous variables. The connection between the frequency contents of the sampled signal and that of the continuous one is given by the equation

$$X_T(\omega) = \frac{1}{T} \sum_{k \in \mathbf{Z}} \hat{x}\left(\omega - \frac{2k\pi}{T}\right) \,. \tag{9.6}$$

This is a standard result from signal analysis, and we refer to the literature for the proof, see for example [16, 22, 23, 28]. The result (9.6) shows that the frequencies outside the interval $[-\pi/T, \pi/T]$ in the analog signal are translated into this interval. This is the aliasing effect of sampling.

We see from (9.6) that the frequency contents of the sampled and the analog real signal will agree, if the nonzero frequencies are in the interval $[-\pi/T, \pi/T]$. If the nonzero frequencies of the analog signal lie in another interval of length $2\pi/T$, then we can assign this interval as the frequency interval of the sampled signal.

The aliasing effect is illustrated in Fig. 9.2. It is also known from everyday life, for example when the wheels of a car turn slowly on a film, although the car is traveling at high speed.

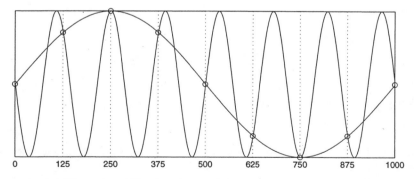

Fig. 9.2. A 7 Hz and a 1 Hz signal sampled 8 times per second yield the same (sampled) signal

9.2 Definition of the Time-Frequency Plane

We use a time-frequency plane to describe how the energy in a signal is distributed with respect to the time and frequency variables. We start with a discrete real signal $\mathbf{x} \in \ell^2(\mathbf{Z})$, with time unit T. We choose $[0, \pi/T]$ as the frequency interval. We mark the sample times on the horizontal axis and the frequency interval on the vertical axis. The sample $x[n]$ contributes $|x[n]|^2$ to the energy, and we place the value in a box located as shown in Fig. 9.3. Alternatively, we fix a grey scale and color the boxes according to the energy contents. This gives a visual representation of the energy!distribution in the signal (see Fig. 9.9).

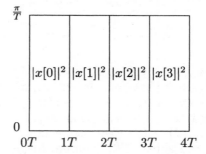

Fig. 9.3. Time-frequency plane for a discrete signal

Suppose now that we down sample the signal $\mathbf{x}$ by two. Then the sampling rate is $1/2T$, and we choose $[0, \pi/2T]$ as the frequency interval. Since we have fixed the units, we get the visual representation shown in Fig. 9.4.

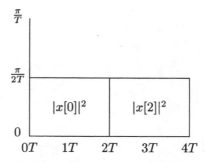

Fig. 9.4. Time-frequency plane for a discrete signal. The signal from Fig. 9.3, down sampled by two

We will now define the time-frequency planes used to visualize the DWT. Let us go back to the first example in Chap. 2. We had eight samples, which we transformed three times using the Haar transform. We first use symbols, and then the numbers from the example. The original signal is represented by eight vertical boxes, like the four vertical boxes in Fig. 9.3. We will take $T = 1$ to simplify the following figures. The first application of the transform is in symbols given as

$$s_3[0], s_3[1], s_3[2], s_3[3], s_3[4], s_3[5], s_3[6], s_3[7]$$
$$\rightarrow s_2[0], s_2[1], s_2[2], s_2[3], d_2[0], d_2[1], d_2[2], d_2[3] \; .$$

Each of the down sampled components, s_2 and d_2, can be visualized as in Fig. 9.4, but not in the same figure, since they both occupy the lower half of the time-frequency plane.

The problem is solved by looking at one step of the DWT in the frequency representation, as described in Sect. 7.3 in the form of a two channel filter bank. We have illustrated the process in Fig. 9.5. The original signal is shown on the left hand side in the second row. Its Fourier transform is shown in the top graph, together with the transfer functions of the two filters. The bottom parts can be obtained in two different ways. In the time domain we use convolution with the filters followed by down sampling by two. In the frequency domain we take the product of the Fourier transform of the signal, $X(\omega)$, and the two transfer functions $H(\omega)$ and $G(\omega)$, and then take the inverse Fourier transform of each product, followed by down sampling by two. All of this is shown in the right part of the figure.

Let us now return to the example with eight samples. The original signal contains frequencies in the interval $[0, \pi]$ (recall that we have chosen $T = 1$). For the s_2 part we can choose the interval $[0, \pi/2]$ and for the d_2 the interval $[\pi/2, \pi]$. If $\mathbf{h}$ is an ideal low pass filter and $\mathbf{g}$ an ideal high pass filter, this gives the correct frequency contents of the two signals. But for real world filters this is only approximately correct. With this choice, and we emphasize that it is a choice, we get the visualization of the transformed signal shown in Fig. 9.6. Usually we use a grey scale to represent the relative values instead of the actual values.

The next step is to apply the DWT to the signal s_2, to obtain s_1 and d_1, each of length two. We assign the frequency interval $[0, \pi/4]$ to s_1, and $[\pi/4, \pi/2]$ to d_1. Thus the two step DWT is visualized as in Fig. 9.7

In the final step s_1 is transformed to s_0 and d_0, each of length one. This third step in the DWT is visualized in Fig. 9.8.

We now illustrate this decomposition with the numbers from Chap. 2. All the information has been gathered in Fig. 9.9. The boxes have been labeled with the coefficients, not their squares, to make it easier to recognize the coefficients. The squares have been used in coloring the boxes.

In Chap. 8 we generalized the wavelet analysis to the wavelet packet analysis. We applied the DWT repeatedly to all elements, down to a given level J, to get a full wavelet packet decomposition to this level, see Fig. 8.2(b) for

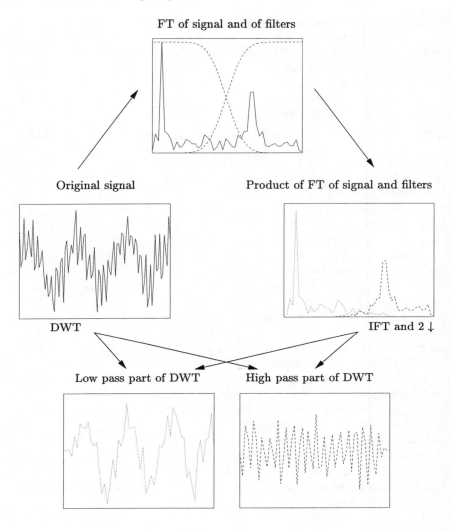

Fig. 9.5. The wavelet transform using the filters **h** and **g** in the frequency domain

$J = 4$. Based on this decomposition, a very large number of different representations of a signal could be obtained. We can visualize these bases using the same idea as for the wavelet decomposition. Each time we apply the DWT to a signal, we should assign frequency intervals as above, assigning the lower interval to the low pass filtered part, and the upper interval to the high pass filtered part. The process leads to partitions of the time-frequency plane, one partition for each possible wavelet packet decomposition. Each partition is a way of visualizing the effect (on any signal) in time and frequency of a chosen

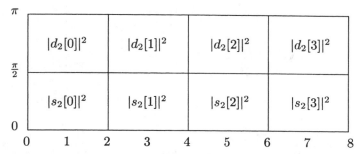

Fig. 9.6. One step DWT applied to eight samples, with $T = 1$. Visualization of energy distribution

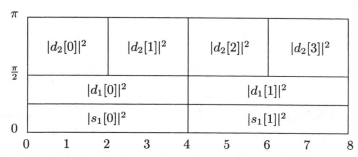

Fig. 9.7. Two step DWT applied to eight samples, with $T = 1$. Visualization of energy distribution

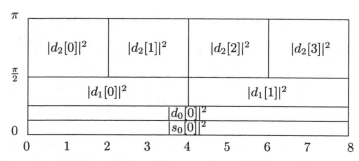

Fig. 9.8. Three step DWT applied to eight samples, with $T = 1$. Visualization of energy distribution

(1)	56	40	8	24	48	48	40	16
(2)	48	16	48	28	8	−8	0	12
(3)	32	38	16	10	8	−8	0	12
(4)	35	−3	16	10	8	−8	0	12

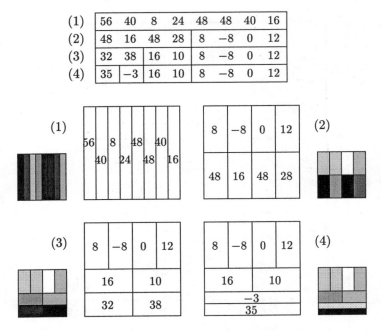

Fig. 9.9. Each level in a wavelet decomposition corresponds to a time-frequency plane. Values from Chap. 2 are used in this example. Boxes are marked with coefficients, not with their squares

representation. For a given signal the boxes can then be colored according to energy contents in each box. This way it becomes easy visually to identify a representation, where the transformed signal has few large coefficients.

The partition of the time-frequency plane associated with a chosen wavelet packet representation can be constructed as above. A different way of obtaining it is to turn the decomposition 90 degrees counterclockwise. The partition then becomes easy to identify. An example with a signal of length 32, decomposed over $J = 4$ levels, is shown in Fig. 9.10.

In Fig. 9.11 we have shown four possible representations for a signal with 8 samples, and the associated partitions of the time-frequency plane.

One more thing concerning the time-frequency planes should be mentioned. The linear scale for energy used up to now is often not the relevant one in applications. One should choose an appropriate grey scale for coloring the cells. Often one uses the logarithm of the energy. Thus the appropriate measure is often $20\log_{10}(|s_j[k]|)$ (units decibel (dB)), displayed using a linear grey scale. In the following figures we have used a logarithmic scale, slightly modified, to avoid that coefficients with values close to zero lead to a compression of the linear grey scale used to display these values. Thus we use

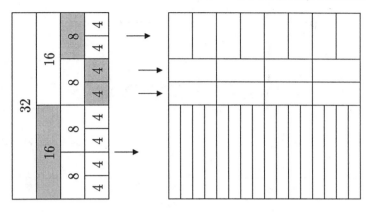

Fig. 9.10. A time-frequency plane is easily constructed by turning the decomposition. The length of each element shows the height of each row of cells, and the number of coefficients in each element the number of cells in each row. The numbers in the boxes are the lengths of the elements

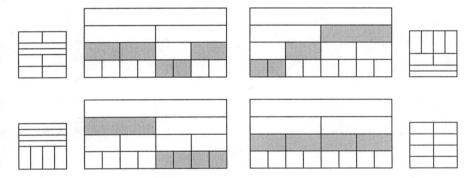

Fig. 9.11. Four different choices of bases and the corresponding time-frequency planes. Note that the figure is valid only for signals of length 8, since there are 8 cells

$\log(1 + |s_j[k]|^2)$ to determine our coloring of the time-frequency planes in the sequel, in combination with a linear grey scale.

9.3 Wavelet Packets and Frequency Contents

We will now take a closer look at the frequency contents in the time-frequency visualization of a wavelet packet analysis, for example those shown in Fig. 9.11. The periodicity of $X_T(\omega)$ and the symmetry property $|X_T(\omega)| = |X_T(-\omega)|$ together imply that when we take for example the frequency interval $[\pi/T, 2\pi/T]$ to determine $X_T(\omega)$, then the frequency contents

108   9. The Time-Frequency Plane

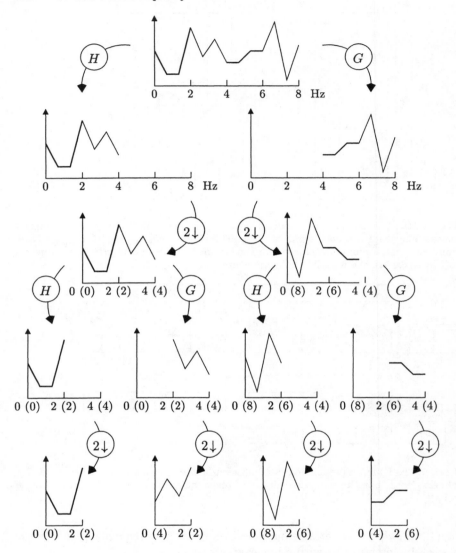

Fig. 9.12. Due to the down sampling all the high pass parts are mirrored. This swaps the low and high pass part in a subsequent transform. The result is that the frequency order of the four signals at the bottom level is 1, 2, 4, and 3. The numbers in parentheses show the origin of parts of the signal. The differences in line thickness is a help to trace the in signal parts. Note that the figure is based on ideal filters

in this interval is the mirror image of the contents in $[0, \pi/T]$, see Fig. 9.1. Thus we have to be careful in interpreting the frequency contents in a wavelet packet analysis. In the first step from $\mathbf{s}_j$ to $\mathbf{s}_{j-1}$, $\mathbf{d}_{j-1}$ we have assigned the interval $[0, \pi/2T]$ to $\mathbf{s}_{j-1}$ and the interval $[\pi/2T, \pi/T]$ to $\mathbf{d}_{j-1}$, in our construction of the time-frequency plane. In the wavelet analysis we only decompose $\mathbf{s}_{j-1}$ in the next step. Here the frequency interval is the one we expect when applying low pass and high pass filters. In the wavelet packet analysis we apply the filters also to the part $\mathbf{d}_{j-1}$. When we apply filters to this part, the frequency interval has to be $[0, \pi/2T]$. But the frequency contents in this interval is the mirror image of the contents in $[\pi/2T, \pi/T]$, which means that the low frequency and high frequency parts are reversed. Two steps in a wavelet packet decomposition are shown in Fig. 9.12.

It is important to understand the frequency ordering of the elements in a wavelet packet analysis. A more extensive example is given in Fig. 9.13. We have taken a 128 Hz signal, which has been sampled, such that the frequency range is from 0 Hz to 64 Hz. Three steps of the full wavelet packet decomposition are shown, such that the figure shows a four level decomposition. The elements at the bottom level have a frequency ordering 0, 1, 3, 2, 6,

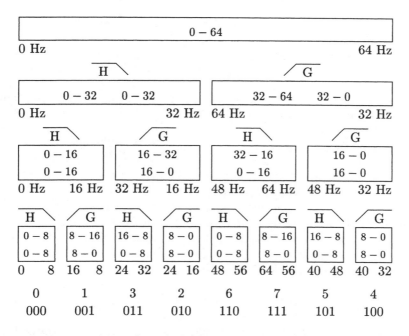

Fig. 9.13. decomposition of a signal following the same principle as in Fig. 9.12. The numbers in the cells shows the real frequency content before (in parentheses) and after down sampling, while the numbers below the cells shows from which frequency band the signal part originates. The last two lines show the frequency order, in decimal and binary notation, respectively

7, 5, 4. We write these numbers in binary notation, using three digits: 000, 001, 011, 010, 110, 111, 101, 100. Then we note a special property of this sequence. Exactly one binary digit changes when we go from one number to the next. It is a special permutation of the numbers 0 to $2^N - 1$. Such a sequence is said to be *Gray code permuted*. The Gray code permutation is formally defined as follows. Given an integer n, write it in binary notation as $n_{N+1}n_N n_{N-1} \ldots n_2 n_1$, such that each n_i is either 0 or 1. Note that we have added a leading zero. It is convenient in the definition to follow. For example, for $n = 6$ we have $n_1 = 0$, $n_2 = 1$, $n_3 = 1$, and $n_4 = 0$. The Gray code permuted integer $GC(n)$ is then defined via its binary representation. The i'th binary digit is denoted by $GC(n)_i$, and is given by the formula

$$GC(n)_i = n_i + n_{i+1} \mod 2 . \qquad (9.7)$$

The inverse can be found, again in binary representation, via the following formula

$$IGC(n)_i = \sum_{k \geq i} n_k \mod 2 . \qquad (9.8)$$

The sum is actually finite, since for an integer $n < 2^N$ we have $n_k = 0$ for all $k > N$.

With these definitions we see that we get from the frequency ordering in Fig. 9.13 to the natural (monotonically increasing) frequency order by using the IGC map.

Once we have seen how the permutation arises in our scheme for finding the wavelet packet decomposition, we can devise the following simple change to the scheme to ensure that the elements appear in the natural frequency order. Above we saw that after one step of the DWT, due to the down sampling, the frequency contents in the high pass part appeared in reverse order (see Fig. 9.12). Thus in the next application of the DWT to this part, the low and high frequency parts appear in reverse order, see again Fig. 9.12. This means that to get the elements in the natural frequency order, we have to interchange the position of the low and high frequency filtered parts in every other application of the DWT step. This method is demonstrated in Fig. 9.14. Look carefully at the figure, notice where the H and G filters are applied, and compare in detail with Fig. 9.13.

Thus we have two possible ways to order the frequency contents in the elements of a full wavelet packet decomposition of a given signal. Using the original scheme we get the ordering as shown in Fig. 9.13. It is called *filter bank ordering*. The other ordering is called *natural frequency ordering*. Sometimes it is important to get the frequency ordering right. In particular, if one wants to interpret the time-frequency plane, then this is important. In other cases, for example in applications to denoising and compression, the ordering is of no importance.

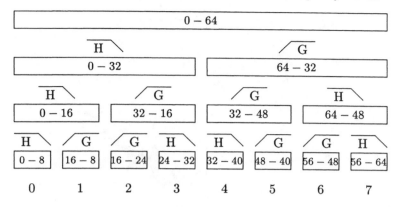

Fig. 9.14. We get the natural frequency order by swapping every other application of the DWT. Compare this figure with Fig. 9.13

Let us illustrate the consequences of the choice of frequency ordering in the time-frequency plane. We take the signal which is obtained from sampling the function $\sin(128\pi t^2)$ in 1024 points in the time interval $[0, 2]$. This signal is called a linear chirp, since the instantaneous frequency grows linearly with time. As the DWT in this example we take the Daubechies 4 transform. Some of the possible bases in a wavelet packet decomposition are those that we call level bases, meaning that we choose all elements in the basis from one fixed level. With a level basis with $J = 6$ we then get the two time-frequency planes shown in Fig. 9.15. Each plane consists of 32×32 cells, colored with a linear grey scale map according to the values of $\log(1 + |s_j[k]|^2)$.

9.4 More about Time-Frequency Planes

We will now discuss in some further detail the time-frequency planes. It is clear from Fig. 9.15 that some improvements can be made. We will divide the discussion into the following topics.

- frequency localization
- time localization
- Alignment
- Choice of basis

Each topic is discussed in a separate subsection.

9.4.1 Frequency Localization

Before we explain the effects in Fig. 9.15 we look at a simpler example. Let us first explain why we choose a level basis in this figure. It is due to the uneven

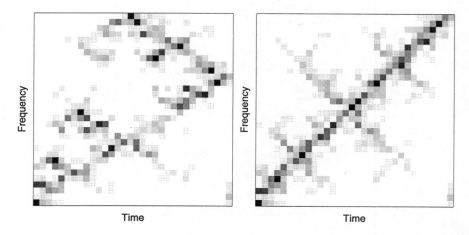

Fig. 9.15. Visualization of significance of frequency ordering of the elements in a decomposition. The left hand plot uses the filter bank ordering, whereas the right hand plot uses the natural frequency order

time-frequency localization properties in a wavelet basis (low frequencies are well localized, and high frequencies are poorly localized), see for example Fig. 9.8, or the fourth example in Fig. 9.11. With a level basis the time-frequency plane is divided into rectangles of equal size. We illustrate this difference with the following example. As the DWT we take the one based on Daubechies 4. As the signal we take

$$\sin(\omega_0 t) + \sin(2\omega_0 t) + \sin(3\omega_0 t) \,, \tag{9.9}$$

sampled at 1024 points in the time interval $[0, 1]$. We have taken $\omega_0 = 405.5419$ to get frequencies that cannot be localized in just one frequency interval, in the partition into 32 intervals in our level basis case. In Fig. 9.16 the frequency plane for the wavelet transform is shown on the left. On the right is a level basis decomposition. In both cases we have decomposed to $J = 6$. For the level basis we then get 32×32 boxes.

It is evident from this figure that the level basis is much better at localizing frequencies in a signal. So in the remainder of this section we use only a level basis.

Let us look again at the right hand part of Fig. 9.15. The signal is obtained by sampling a linear chirp. Thus we expect the frequency contents to grow linearly with time. This is indeed the main impression one gets from the figure. But there are also reflections of this linear dependence. The reflections appear at frequencies $f_s/4, f_s/8, \dots$, where f_s is the sampling frequency. This is a visualization of the reflection in frequency contents due to down sampling, as discussed in the previous section. It is due to our use of real filters instead of ideal filters.

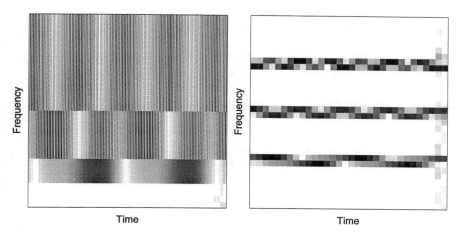

Fig. 9.16. A signal with three frequencies in the time-frequency plane. Left hand plot shows the time-frequency plane for the wavelet decomposition, and the right hand plot the level basis from a wavelet packet decomposition, both with $J = 6$ frequency localization

We therefore start by looking at the frequency response of some wavelet filters. In Sect. 7.3 we showed the frequency response for the filters for the three transforms, which we call Haar (also called Daubechies 2), in Fig. 7.5, Daubechies 4, in Fig. 7.6, and CDF(2,2), in Fig. 7.7. Note that the scale on the vertical axis is linear in these three figures. It is more common to use a logarithmic scale. Let us show a number of plots using a logarithmic scale. In Fig. 9.17 the left hand part shows the frequency response of the filters Daubechies 2, 12, and 22. The right hand part shows the frequency response of CDF(4,6). All figures show that the filters are far from ideal. By increasing the length of the Daubechies filters one can get closer to ideal filters. In the limit they become ideal.

In Fig. 9.18 we have repeated the plot from the right hand part of Fig. 9.15, which was based on Daubechies 4, and then also plotted the same time-frequency plane, but now computed using Daubechies 12 as the DWT. It is evident that the sharper filter gives rise to less prominent reflections in the time-frequency plane.

The right hand plot in Fig. 9.17 shows another problem that may occur. The frequency response of the low pass and the high pass parts of CDF(4,6) is not symmetric around the normalized frequency 0.5, i.e. the frequency divided by π. In repeated applications of the DWT this asymmetry leads to problems with the interpretation of the frequency contents. We will illustrate this with an example. Suppose that we apply the CDF(4,6) transforms three times to get to the fourth level in a full wavelet packet decomposition. Thus we have eight elements on the fourth level, as shown in Fig. 8.2(b). We can

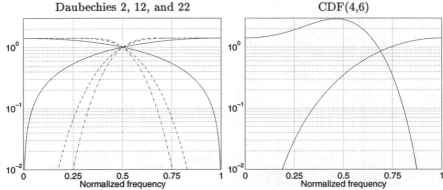

Fig. 9.17. Frequency response for Daubechies 2, 12 (*dashed*), and 22 (*dasheddotted*), and for the biorthogonal CDF(4,6)

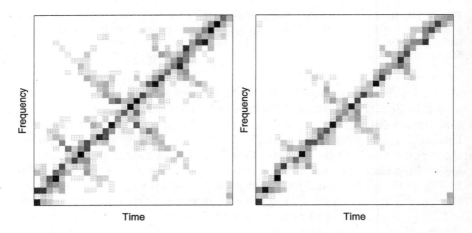

Fig. 9.18. The left hand part is the time-frequency plane for the linear chirp from Fig. 9.15, which is based on Daubechies 4. The same plot, based on Daubechies 12, is shown in the right hand part

find the frequency response of the corresponding eight bandpass filters. They are plotted in Fig. 9.19

Let us also illustrate the reflection of the line in a linear chirp due to undersampling. If we increase the frequency range beyond the maximum given by the sampling frequency, then we get a reflection as shown in Fig. 9.20. This figure is based on Daubechies 12 filters.

9.4.2 Time Localization

It is easy to understand the time localization properties of the DWT step, using the filter bank approach. We only consider FIR filters. In the time

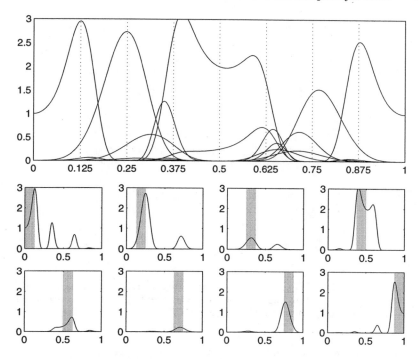

Fig. 9.19. The eight bandpass filters corresponding to the fourth level in a decomposition based on CDF(4,6). The top part shows the eight plots together, and the bottom part shows the individual responses. The ideal filter response is shaded on each of these figures

domain the filter acts by convolution. Suppose that $\mathbf{h} = \big[h[1], h[2], \dots, h[N]\big]$ is a filter of length N. Then filtering the signal $\mathbf{x}$ yields

$$(\mathbf{h} * \mathbf{x})[n] = \sum_{k=1}^{N} h[k]x[n-k] \; . \tag{9.10}$$

Thus the computed value at n only depends on the preceding samples $x[n-1], \dots, x[n-N]$.

Let us illustrate the time localization properties of the wavelet decomposition and the best level decomposition, both with $J = 6$, i.e. five applications of the DWT. We use Daubechies 4 again, and take the following signal of length 1024.

$$x[n] = \begin{cases} 25 & \text{if } n = 300 \; , \\ 1 & \text{if } 500 \le n \le 700 \; , \\ 15 & \text{if } n = 900 \; , \\ 0 & \text{otherwise} \; . \end{cases} \tag{9.11}$$

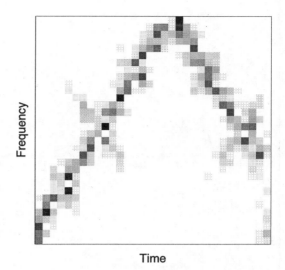

Frequency

Time

Fig. 9.20. This time-frequency plane shows the effect of undersampling a linear chirp. Above the sampling rate the frequency contents is reflected into the lower range. This plot uses Daubechies 12 as the DWT

The two time-frequency planes are shown in Fig. 9.21. We notice that the wavelet transform is very good at localizing the singularities and the constant part. The wavelet packet best level representation has much less resolution with respect to time. The filter is so short (length 4) that the effects of the filter length is not very strong. We see the expected broadening in the wavelet transform of the contribution from the singularity, due to the repeated applications of the filters. You should compare this figure with Fig. 4.6.

Let us give one more illustration. This time we take the sum of the signals used in Fig. 9.16 and Fig. 9.21. The two plots are shown in Fig. 9.22. Here we can see that the wavelet packet best basis representation gives a reasonable compromise between resolution in time and in frequency.

9.4.3 Alignment

In Fig. 9.23 we have plotted the impulse response (filter coefficients) of the filters Daubechies 24 and Coiflet 24 (in the literature also called coif4), both of length 24. The first column shows the IR of the low pass filters, and the second column those of the high pass filters. We see that only a few coefficients dominate. The Coiflet is much closer to being symmetric, which is significant in applications, since it is better at preserving time localization. Let us explain this is some detail.

It is evident from Fig. 9.23 that the large coefficients in the filters can be located far from the middle of the filter. If we recall from Chap. 7 that with

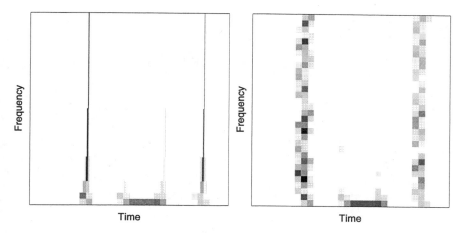

Fig. 9.21. The left hand plot shows the time-frequency plane for the signal in (9.11), decomposed using the wavelet transform, and the right hand plot the same signal in a level basis decomposition, both with $J = 6$

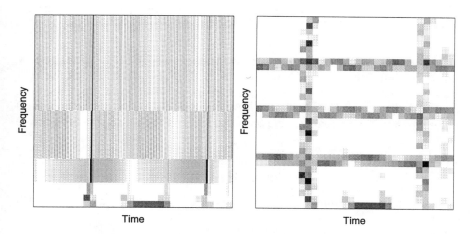

Fig. 9.22. The left hand plot shows the time-frequency plane for the signal in (9.11) plus the one in (9.9), decomposed using the wavelet transform, and the right hand plot the same signal in a level basis decomposition, both with $J = 6$

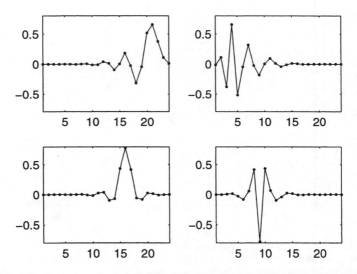

Fig. 9.23. The first row shows the IR of the Daubechies 24 filters. The second row shows the same plots for Coiflet 24. The left hand column shows the IR of the low pass filters, the right hand one those of the high pass filters

orthogonal filters the high pass filter is obtained from the low pass filter by reflection and alternation of the signs, see (7.68), then the center of the high pass filter will be in the opposite end. This is also clear from Fig. 9.23.

For a filter $\mathbf{h} = [h[1], h[2], \ldots, h[N]]$ we can define its center in several different ways. For a real number x we let $\lfloor x \rfloor$ denote the largest integer less than or equal to x.

Maxima location The center is defined to be the first occurrence of the absolute maximum of the filter coefficients. Formally this is defined by

$$C_{\max}(\mathbf{h}) = \min\{n \mid |h[n]| = \max\{|h[k]| \mid k = 1, \ldots, N\}\} . \quad (9.12)$$

Mass center The mass center is defined by

$$C_{\text{mass}}(\mathbf{h}) = \left\lfloor \frac{\sum_{k=1}^{N} k|h[k]|}{\sum_{k=1}^{N} |h[k]|} \right\rfloor . \quad (9.13)$$

Energy center The energy center is defined by

$$C_{\text{energy}}(\mathbf{h}) = \left\lfloor \frac{\sum_{k=1}^{N} k|h[k]|^2}{\sum_{k=1}^{N} |h[k]|^2} \right\rfloor . \quad (9.14)$$

As an example the values for the filters in Fig. 9.23 are shown in Table 9.1.

Suppose now that the signal $\mathbf{x}$, which is being filtered by convolution by $\mathbf{h}$ and $\mathbf{g}$, only has a few large entries. Then these will be shifted in location by

Table 9.1. Centers of Daubechies 24 and Coiflet 24

Filter	$C_{\max}$	C_{mass}	C_{energy}
Daub24 **h**	21	19	20
Daub24 **g**	4	5	4
Coif24 **h**	16	15	15
Coif24 **g**	9	9	9

the filtering, and the shifts will differ in the low and high pass filtered parts. In the full wavelet packet decomposition this leads to serious misalignment of the various parts. Thus shifts have to be introduced. As an example, we have again used the linear chirp, and the Daubechies 12 filters. The left hand part of Fig. 9.24 shows the time-frequency plane based on the unaligned level decomposition, whereas the right hand part shows the effect on alignment based on shifts computed using $C_{\max}$.

Alignment based on the three methods for computing centers given above is implemented in *Uvi_Wave*. As can be guessed from Fig. 9.24, we have used alignment in the other time-frequency plane plots given in this chapter. We have used $C_{\max}$ to compute the alignments. The various possibilities are selected using the function wtmethod in *Uvi_Wave*.

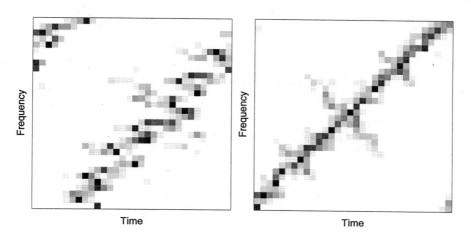

Fig. 9.24. The left hand part shows the time-frequency plane for the linear chirp in Fig. 9.18 without alignment corrections. The right hand part is repeated from this figure

9.4.4 Choice of Basis

It is evident from both Fig. 9.21 and Fig. 9.22 that the choice of basis determines what kind of representation one gets in the time-frequency plane. There are basically two possibilities. One can decide from the beginning that one is to use a particular type of bases, for example a level basis, and then plot time-frequency planes for a signal based on this choice. As the above examples show this can be a good choice. In other cases the signal may be completely unknown, and then the best choice may be to use a particular cost function and find the best basis relative to this cost function. The time-frequency plane is then based on this particular basis. As an example we have taken the signal used in Fig. 9.22 and found the best basis using Shannon entropy as the cost function. The resulting time-frequency plane is shown in Fig. 9.25. One should compare the three time-frequency planes in Fig. 9.22 and Fig. 9.25. It is not evident which is the best representation to determine the time-frequency contents of this very simple signal.

The simple examples given here show that to investigate the time-frequency contents of a given signal may require the plot of many time-frequency planes. In the best basis algorithm one may have to try several different cost functions, or for example different values of the parameter p in the ℓ^p-norm cost function.

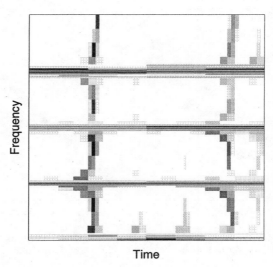

Fig. 9.25. The time-frequency plane for the signal from Fig. 9.22, in the best basis determined using Shannon entropy as the cost function

9.5 More Fourier Analysis. The Spectrogram

We now present a different way to visualize the distribution of the energy with respect to time and frequency. It is based on the short time Fourier transform, and in practical implementations, on the discrete Fourier transform. The resulting visualization is based on the spectrogram. To define it we need some preparation.

Given a real signal $\mathbf{x} \in \ell^2(\mathbf{Z})$, and a sampling rate $1/T$, we have visualized the energy distribution as in Fig. 9.3. Here we have maximal resolution with respect to time, since we take each sample individually. But we have no frequency information beyond the size of the frequency interval, which is determined by the sampling rate. On the other hand, if we use all samples, then we can compute $X_T(\omega)$, which gives detailed information on the distribution of the energy with respect to frequency. One can interpret $\frac{T}{2\pi}|X_T(\omega)|^2$ as the energy density, as can be seen from Parseval's equation (9.3). The energy in a given frequency interval is obtained by integrating this density over that interval.

We would like to make a compromise between the two approaches. This is done in the short time Fourier transform. One chooses a window vector $\mathbf{w} = \{w[n]\}_{n \in \mathbf{Z}}$, which is a sequence with the property that $0 \leq w[n] \leq 1$ for all $n \in \mathbf{Z}$. Usually one chooses a window with only a finite number of nonzero entries. In Fig. 9.26 we have shown four typical choices, each with 16 nonzero entries.

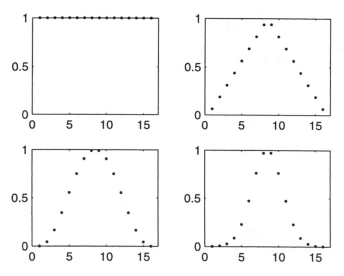

Fig. 9.26. Four window vectors of length 16. Top row shows rectangular and triangular windows. Bottom row shows the Hanning window on the left and a Gaussian window on the right

Once a window vector is chosen, a short time Fourier transform of a signal $\mathbf{x}$ is computed as

$$X_{\mathrm{STFT}}(k,\omega) = \sum_{n \in \mathbf{Z}} w[n-k]x[n]e^{-jnT\omega} . \qquad (9.15)$$

The window is moved to position k and then one computes the Fourier series of the sequence $w[n-k]x[n]$, which is localized to this window. This is repeated for values of k suitably spaced. Suppose that the length N of the window is even. Then one usually chooses $k = mN/2$, $m \in \mathbf{Z}$. For N odd one can take $k = m(N-1)/2$. Thus ones slides the window over the signal and looks at the frequency contents in each window.

The function $|X_{\mathrm{STFT}}(k,\omega)|^2$ is called a spectrogram. It shows the energy density (or power) distribution in the signal, based on the choice of window vector.

Take as the window vector the constant vector

$$w[n] = 1 \quad \text{for all } n \in \mathbf{Z} ,$$

then for all k we have $X_{\mathrm{STFT}}(k,\omega) = X_T(\omega)$, the usual Fourier series. On the other hand, if one chooses the shortest possible rectangular window,

$$w[n] = \begin{cases} 1 & \text{for } n = 0 , \\ 0 & \text{for } n \neq 0 , \end{cases}$$

then one finds $X_{\mathrm{STFT}}(k,\omega) = x[k]e^{-jkT\omega}$, such that the time-frequency plane in Fig. 9.3 gives a visualization of the spectrogram for this choice of window.

Concerning the window vectors in Fig. 9.26, then the rectangular one of length N is given by $w[k] = 1$ for $k = 1, \ldots, N$, and $w[k] = 0$ otherwise. The triangular window is defined for N odd by

$$w[n] = \begin{cases} 2n/(N+1), & 1 \leq n \leq (N+1)/2 , \\ 2(N-n+1)/(N+1), & (N+1)/2 \leq n \leq N , \end{cases}$$

and for N even by

$$w[n] = \begin{cases} (2n-1)/N, & 1 \leq n \leq N/2 , \\ 2(N-n+1)/N, & N/2 \leq n \leq N . \end{cases}$$

All values $w[n]$ not defined by these equations are zero. The Hanning (or Hann) window of length N is defined by

$$w[n] = \sin^2(\pi(n-1)/N), \quad n = 1, \ldots, N .$$

This window is often used for the short time Fourier transform. The last window in Fig. 9.26 is a Gaussian window, defined by

$$w[n] = \exp(-\sigma(2n - (N+1))^2), \quad n = 1, \ldots, N \, ,$$

for a positive parameter σ. In our example $\sigma = 0.03$. Note that the four window vectors can be obtained by sampling a box function, a hat function, $\sin^2(\pi t)$, and $\exp(-\sigma t^2)$, respectively.

The above results can be applied to a finite signal, but for such signals one usually chooses a different approach, based on the discrete Fourier transform, here abbreviated as DFT. Since the Fourier expansion involves complex numbers, it is natural to start with complex signals, although we very soon restrict ourselves to real ones. A finite signal of length N will be indexed by $n = 0, \ldots, N - 1$. All finite signals of length N constitute the vector space $\mathbf{C}^N$, of dimension N. We define

$$e_k[n] = e^{j2\pi kn/N}, \quad n, k = 0, \ldots, N - 1 \, . \tag{9.16}$$

The vectors e_k are orthogonal with respect to the usual inner product on $\mathbf{C}^N$, see (7.44). Thus these vectors form a basis for the space of finite signals $\mathbf{C}^N$. The DFT is then expansion of signals with respect to this basis. We use the notation $\hat{\mathbf{x}}$ for the DFT of $\mathbf{x}$. The coefficients are given by

$$\hat{x}[k] = \sum_{n=0}^{N-1} x[n] e^{-j2\pi kn/N} \, . \tag{9.17}$$

The inversion formula is

$$x[n] = \frac{1}{N} \sum_{k=0}^{N-1} \hat{x}[k] e^{j2\pi kn/N} \, . \tag{9.18}$$

Parseval's equation is for the DFT the equation

$$\|\mathbf{x}\|^2 = \frac{1}{N} \sum_{k=0}^{N-1} |\hat{x}[k]|^2 \, . \tag{9.19}$$

Let us note that the computations in (9.17) and (9.18) have fast implementations, known as the fast Fourier transform.

Comparing the definition of the DFT (9.17) with the definition of the Fourier series (9.1), we see that

$$\hat{x}[k] = X(2\pi k/N) \, .$$

Thus the DFT can be viewed as a sampled version of the Fourier series, with the sample points $0, 2\pi/N, \ldots, 2\pi(N-1)/N$.

For a real $\mathbf{x}$ of length N we have

$$\overline{\hat{x}}[k] = \sum_{n=0}^{N-1} x[n] e^{j2\pi nk/N} = \sum_{n=0}^{N-1} x[n] e^{-j2\pi n(N-k)/N} = \hat{x}[N-k] \, , \tag{9.20}$$

where we introduce the convention that $\hat{x}[N] = \hat{x}[0]$. This is consistent with (9.17). This formula allows us to define $\hat{x}[k]$ for any integer. It is then periodic with period N.

We see from (9.20)that we need only half of the DFT coefficients to reconstruct $\mathbf{x}$, when the signal is real.

Now the spectrogram used in signal processing, and implemented in the signal processing toolbox for MATLAB as specgram, is based on the DFT. A window vector $\mathbf{w}$ is chosen, and then one computes $|X_{\mathrm{STFT}}(k, 2\pi n/N)|^2/2\pi$ for values of k determined by the length of the window vector and for $n = 0, 1, \ldots, N - 1$. The spectrogram is visualized as a time-frequency plane, where the cells are chosen as follows. Along the time axis the number of cells is determined by the length of the signal N, the length of the window vector, and the amount of overlap one wishes to use. The number of cells in the direction of the frequency axis is determined by the length of the window (the default in specgram). Assume the length of the window is L, and as usual that the signal is real. If L is even, then there will be $(L/2) + 1$ cells on the frequency axis, and if L is odd, there will be $(L + 1)/2$ cells. If the sampling rate is know, then it is used to determine the units on the frequency axis.

Let us give two spectrograms of the signal used in Fig. 9.22, and also in Fig. 9.25. They are obtained using the specgram function from the signal processing toolbox for MATLAB. In the plot on the left hand side of Fig. 9.27 a Hanning window of length 256 is used. In the right hand plot the window length is 64. In both plots we have used a color map which emphasizes the large values. Larger values are darker.

The trade-off between time and frequency localization is clearly evident in these two figures. One should also compare this figure with Fig. 9.22 and with Fig. 9.25. Together they exemplify both the possibilities and the complexity in the use of time-frequency planes to analyze a signal.

9.5.1 An Example of Fourier and Wavelet Time-Frequency Analysis

To wrap up this chapter we will give a more complex example of an application of the best basis algorithm. We have taken a signal consisting of several different signals added, and made two wavelet and two Fourier time-frequency analyses of the signal.

In Fig. 9.28 the signal is shown in the four different time-frequency planes. The first two are the traditional spectrograms based on the STFT. Here we must choose between long and short oscillations. With a short window we see the short oscillations clearly, but long oscillations becomes inaccurate with respect to frequency, while the long window tends to smear the short oscillation and enhance the long oscillations. Using a wavelet time-frequency plane with equally dimensions cells (level basis) does improve the time-frequency plane, primarily due to the very short filter, that is length 12 compare to the

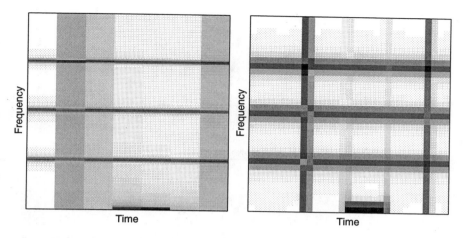

Fig. 9.27. The left hand part shows the time-frequency plane for the signal from Fig. 9.22 obtained as a spectrogram with a Hanning window of length 256. In the right hand part the window length is 64. The signal has 1024 samples

shortest window of length 64 in the STFT. But when using the best basis algorithm to pick a basis, in this case based on Shannon's entropy, the time-frequency plane is much improved. The different sized cells makes it possible to target the long, slow oscillations in the lower half *and* the fast and short oscillations in the upper half.

Exercises

After you have read Chap. 13, you should return to these exercises.

9.1 Start by running the **basis** script in *Uvi_Wave*. Then use the *Uvi_Wave* functions **tfplot** and **tree** to display the basis tree graph and the time-frequency plane tilings for a number of different basis.

9.2 Get the function for coloring the time-frequency tilings from the Web site of this book (see Chap. 1) and reproduce the figures in this chapter.

9.3 Do further computer experiments with the time-frequency plane and synthetic signals.

9.4 If you have access to the signal processing toolbox, do some experiments with the **specgram** function to understand this type of time-frequency plane. Start with simple signals with known frequency contents, and vary the window type and length.

Spectrogram, 1024 point FFT, windows 64, overlap 16

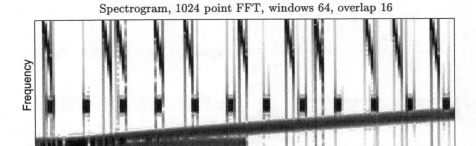

Spectrogram, 1024 point FFT, windows 512, overlap 400

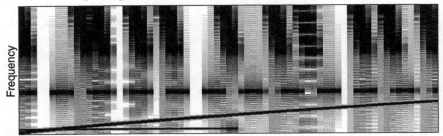

Scalogram with level basis, Symlet 12

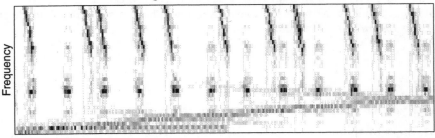

Scalogram with best basis, Symlet 12, and the entropy cost function

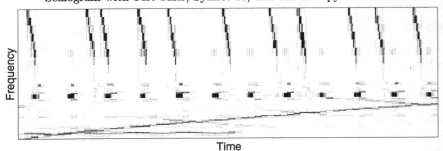

Time

Fig. 9.28. Four different time-frequency planes of the same signal. The signal is a composite test signal consisting of one slow and fourteen very fast chirps, a fixed frequency lasting the first half of the signal, and a periodic sinus-burst. The window used in the Fourier analyses is a Hanning

10. Finite Signals

In the previous chapters we have only briefly, and in a casual way, considered the problems arising from having a finite signal. In the case of the Haar transform there are no problems, since it transforms a signal of even length to two parts, each of half the original length. In the case of infinite length signals there are obviously no problems either. But in other cases we may need for instance sample $s[-1]$, and our given signal starts with sample $s[0]$. We will consider solutions to this problem, which we call the boundary problem. Theoretical aspects are considered in this chapter. It is important to understand that there is no universal solution to the boundary problem. The preferred solution depends on the kind of application one has in mind. The implementations are discussed in Chap. 11. The reader mainly interested in implementations and applications can skip ahead to this chapter.

Note that in this chapter we use a number of results from linear algebra. Standard texts contain the results needed. Note also that in this chapter we use both row vectors and column vectors, and that the distinction between the two is important. The default is column vectors, so we will repeatedly state when a vector is a row vector to avoid confusion. Some of the results in this chapter are established only in an example, and the interested reader is referred to the literature for the general case.

There is a change in notation in this chapter. Up to now we have used the notation $\mathbf{h}_0$, $\mathbf{h}_1$, for the analysis filter pair in the filter bank version of the DWT. From this chapter onwards we change to another common notation, namely $\mathbf{h}$, $\mathbf{g}$ (except for Chap. 12 which is closely connected to Chap. 7). This is done partly to simplify the matrix notation below, partly because the literature on the boundary problem typically uses this notation. We also recall that we only consider filters with real coefficients.

10.1 The Extent of the Boundary Problem

To examine the extent of the problem with finite signals we use the lifting steps for the Haar transform and Daubechies 4. We recall the definition of the Haar transform

$$d^{(1)}[n] = S[2n+1] - S[2n] \,, \tag{10.1}$$

$$s^{(1)}[n] = S[2n] - \frac{1}{2}d^{(1)}[n] \,, \tag{10.2}$$

and of the Daubechies 4 transform

$$s^{(1)}[n] = S[2n] + \sqrt{3}S[2n+1] \,, \tag{10.3}$$

$$d^{(1)}[n] = S[2n+1] - \frac{1}{4}\sqrt{3}s^{(1)}[n] - \frac{1}{4}(\sqrt{3}-2)s^{(1)}[n-1] \,, \tag{10.4}$$

$$s^{(2)}[n] = s^{(1)}[n] - d^{(1)}[n+1] \,. \tag{10.5}$$

Compared to previous equations we have omitted the index j, and the original signal is now denoted by S. When calculating $s^{(1)}[n]$ and $d^{(1)}[n]$ for the Haar transform we need the samples $S[2n]$ and $S[2n+1]$ for each value of n, while for $s^{(1)}[n]$ and $d^{(1)}[n]$ in the case of Daubechies 4 we need $S[2n-2]$, $S[2n-1]$, $S[2n]$, and $S[2n+1]$. The latter is seen by inserting (10.3) into (10.4). For a signal of length 8 the parameter n assumes the values 0, 1, 2, and 3. To perform the Haar transform samples $S[0]$ through $S[7]$ are needed, while the Daubechies 4 transforms requires samples $S[-2]$ through $S[7]$, i.e. the eight known samples and two unknown samples. Longer transforms may need even more unknown samples. This is the boundary problem associated with the wavelet transform.

There exists a number of different solutions to this problem. Common to those we consider is the preservation of the perfect reconstruction property of the wavelet transform. We will explore the three most often used ones, which are *boundary filters*, *periodization*, and *mirroring*. Moreover, we will briefly discuss a more subtle method based on preservation of vanishing moments.

10.1.1 Zero Padding, the Simple Solution

We start with a simple and obvious solution to the problem, which turns out to be rather unattractive. Given a finite signal, we add zeroes before and after the given coefficients to get a signal of infinite length. This is called *zero padding*. In practice this means that when the computation of a coefficient in the transform requires a sample beyond the range of the given samples in the finite signal, we use the value zero.

If we take a signal with 8 samples, and apply zero padding, then we see that in the Haar transform case we can get up to 4 nonzero entries in $\mathbf{s}^{(1)}$ and in $\mathbf{d}^{(1)}$. Going through the steps in the Daubechies 4 transform we see that in $\mathbf{s}^{(1)}$ the entries with indices $0, 1, 2, 3$ can be nonzero, whereas in $\mathbf{d}^{(1)}$ the entries with indices $0, 1, 2, 3, 4$ can be nonzero, and in $\mathbf{s}^{(2)}$ those with indices $-1, 0, 1, 2, 3$ can be nonzero. Thus in the two components in the transform we may end up with a total of 10 nonzero samples.

This is perhaps unexpected, since up to now we have ignored this phenomenon. Previously we have stated that the transform of a signal of even length leads to two components, each of half the length of the input signal. This is correct here, since we have added zeroes, so both the original signal and the two transformed parts have infinite length. For finite signal the statement is only correct, when one uses the Haar transform, or when one applies the right boundary correction. Thus when we use zero padding, the number of nonzero entries will in general increase each time we apply the DWT step.

It is important to note that all 10 coefficients above are needed to reconstruct the original signal, so we cannot just leave out two of them, if the perfect reconstruction property is to be preserved. In general the number of extra coefficients is proportional to the filter length. For orthogonal transforms (such as those in the Daubechies family) the number of extra signal coefficients is exactly $L - 2$, with L being the filter length. See p. 135 for the proof.

When we use zero padding, the growth in the number of nonzero entries is unavoidable. It is not a problem in the theory, but certainly in applications. Suppose we have a signal of length N and a filter of length L, and suppose we want to compute the DWT over k scales, where k is compatible with the length of the signal, i.e. $N \geq 2^k$. Each application of the DWT adds $L - 2$ new nonzero coefficients, in general. Thus the final length of the transformed signal can be up to $N + k(L - 2)$.

The result of using zero padding is illustrated as in Fig. 10.1. As the filter taps "slides" across the signal a number of low and high pass transform coefficients are produced, a pair for each position of the filter. Since there are $(N + L)/2 - 1$ different positions, the total number of transform coefficients is twice this number, that is $N + L - 2$.

If one considers wavelet packet decompositions, then the problem is much worse. Suppose one computes the full wavelet packet decomposition down to a level J, i.e. we apply the DWT building block $J - 1$ times, each time to all elements in the previous level. Starting with a signal of length N and a filter of length L, then at the level J the total length of the transformed signal can be up to $N + (2^{J-1} - 1)(L - 2)$. This exponential growth in J makes zero padding an unattractive solution to the boundary problem.

Thus it is preferable to have available boundary correction methods, such that application of the corrected DWT to a signal leads to two components, each of half the length of the original signal. Furthermore we would like to preserve the perfect reconstruction property. We present four different methods below. The first three methods use a number of results from linear algebra. The fourth method requires extensive knowledge of the classical wavelet theory and some harmonic analysis. It will only be presented briefly and incompletely.

The reader interested in implementation can go directly to Chap. 11.

The methods are presented here using the filter bank formulation of the DWT step. Another solution to the boundary problem based directly on the lifting technique is given in Sect. 11.4.2 with CDF(4,6) as an example. The

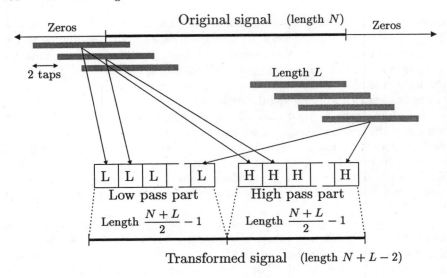

Fig. 10.1. The result of zero padding when transforming a finite signal. The grey boxes illustrates the positions of the filter taps as the filtering occurs. Each position gives a low pass and high pass coefficient. The number of positions determines the number of transform coefficients. In this figure we have omitted most of the 'interior' filters to simplify it

connection between filter banks and lifting is discussed in Chap. 7 and more thoroughly in Chap. 12.

10.2 DWT in Matrix Form

In the previous chapters we have seen the DWT as a transform realized in a sequence of lifting steps. We have also described how the DWT can be performed as a low and high pass filtering, followed by down sampling by 2. Now we turn our attention to the third possibility, which was presented in Chap. 5 using the Haar transform as an example. The transform can be carried out by multiplying the signal with an appropriate matrix. The reconstruction can also be done by a single multiplication. We assume that the input signal, denoted by $\mathbf{x}$, is of even length.

We recall from (7.51) that the low pass filtered and down sampled signal is given as

$$(H\mathbf{x})[n] = \sum_k h[2n - k]x[k] . \qquad (10.6)$$

This convolution is interpreted as an inner product between the vector

$$[\cdots 0 \; h[L-1] \; \cdots \; h[1] \; h[0] \; 0 \; \cdots]$$

and the vector $\mathbf{x}$, or as the matrix product of the reversed filter row vector and the signal column vector. The high pass part $G\mathbf{x}$ is found analogously, see (7.52). The symbols H and G emphasize that we consider the transition from $\mathbf{x}$ to $H\mathbf{x}$ and $G\mathbf{x}$ as a linear maps.

Thus we decompose $\mathbf{x}$ into $H\mathbf{x}$ and $G\mathbf{x}$, and we have to decide how to combine these two components into a single vector, to get the matrix form of the transform. There are two obvious possibilities. One is to take all the components in $H\mathbf{x}$, followed by all components in $G\mathbf{x}$. This is not an easy solution to use, when one considers infinite signals. The other possibility is to interlace the components in a column vector as

$$\mathbf{y} = [\cdots \; (H\mathbf{x})[-1] \; (G\mathbf{x})[-1] \; (H\mathbf{x})[0] \; (G\mathbf{x})[0] \; (H\mathbf{x})[1] \; (G\mathbf{x})[1] \; \cdots]^\top .$$

Since the four vectors in an orthogonal filter set have equal even length (in contrast to most biorthogonal filter sets), it is easier to describe the matrix form of the DWT for orthogonal filters. Later on it is fairly easy to extend the matrix form to biorthogonal filters.

It follows from (10.6) that the matrix of the direct transform has the following structure. The rows of the matrix consist of alternating, reversed low and high pass IRs, each low pass IR is shifted two places in relations to the preceding high pass IR, while the following high pass IR is not shifted. The low pass filter is now denoted by $\mathbf{h}$ and the high pass filter by $\mathbf{g}$, in contrast to Chap. 7, where we used the notation $\mathbf{h}_0$ and $\mathbf{h}_1$, respectively.

If the length of the filter is 6, then the matrix becomes

$$\mathbf{T}_a = \begin{bmatrix} \ddots & \ddots & \ddots & \ddots & \ddots & \ddots & & & & & \\ \ddots & h[5] & h[4] & h[3] & h[2] & h[1] & h[0] & 0 & 0 & 0 & 0 & \cdots \\ & g[5] & g[4] & g[3] & g[2] & g[1] & g[0] & 0 & 0 & 0 & 0 & \cdots \\ \cdots & 0 & 0 & h[5] & h[4] & h[3] & h[2] & h[1] & h[0] & 0 & 0 & \cdots \\ \cdots & 0 & 0 & g[5] & g[4] & g[3] & g[2] & g[1] & g[0] & 0 & 0 & \cdots \\ \cdots & 0 & 0 & 0 & 0 & h[5] & h[4] & h[3] & h[2] & h[1] & h[0] \\ \cdots & 0 & 0 & 0 & 0 & g[5] & g[4] & g[3] & g[2] & g[1] & g[0] & \ddots \\ & & & & & \ddots & \ddots & \ddots & \ddots & \ddots & \ddots \end{bmatrix} . \qquad (10.7)$$

Given an infinite signal $\mathbf{x}$ as a column vector the wavelet transform can be calculated simply by $\mathbf{y} = \mathbf{T}_a\mathbf{x}$. Obviously we want to be able to reconstruct the original signal in the same manner, so we need another matrix such that

$$\mathbf{T}_s\mathbf{T}_a = \mathbf{I} . \qquad (10.8)$$

By multiplying $\mathbf{T}_s$ and $\mathbf{y}$ we get $\mathbf{x}$. For finite matrices the equation (10.8) implies that $\mathbf{T}_s = \mathbf{T}_a^{-1}$, and for infinite matrices we impose this condition. Fortunately, it is easy to show that $\mathbf{T}_a^{-1} = \mathbf{T}_a^\top$ for orthogonal filters (see Exer. 10.2). Recall that a real matrix with this property is called orthogonal. Now we have

$$\mathbf{x} = \mathbf{T}_s \mathbf{T}_a \mathbf{x} = \mathbf{T}_s \mathbf{y} = \mathbf{T}_s \begin{bmatrix} \vdots \\ (H\mathbf{x})[0] \\ (G\mathbf{x})[0] \\ (H\mathbf{x})[1] \\ (G\mathbf{x})[1] \\ (H\mathbf{x})[2] \\ (G\mathbf{x})[2] \\ \vdots \end{bmatrix},$$

so in order to reconstruct the original signal the matrix $\mathbf{T}_s$ is applied to a mix of low and high pass coefficients.

The major difference in the case of biorthogonal filters is that $\mathbf{T}_a$ is not orthogonal, and hence $\mathbf{T}_s$ cannot be found simply by transposing the direct transform matrix. To understand how $\mathbf{T}_s$ is constructed in this case, we first examine $\mathbf{T}_s$ in the orthogonal case. It is easy to show that

$$\mathbf{T}_s = \mathbf{T}_a^\top = \begin{bmatrix} \ddots & \ddots & & \vdots & \vdots & \vdots & \vdots \\ \ddots & \tilde{h}[0] & \tilde{g}[0] & 0 & 0 & 0 & 0 \\ \ddots & \tilde{h}[1] & \tilde{g}[1] & 0 & 0 & 0 & 0 \\ \ddots & \tilde{h}[2] & \tilde{g}[2] & \tilde{h}[0] & \tilde{g}[0] & 0 & 0 \\ \ddots & \tilde{h}[3] & \tilde{g}[3] & \tilde{h}[1] & \tilde{g}[1] & 0 & 0 \\ \ddots & \tilde{h}[4] & \tilde{g}[4] & \tilde{h}[2] & \tilde{g}[2] & \tilde{h}[0] & \tilde{g}[0] \\ & \tilde{h}[5] & \tilde{g}[5] & \tilde{h}[3] & \tilde{g}[3] & \tilde{h}[1] & \tilde{g}[1] & \ddots \\ & 0 & 0 & \tilde{h}[4] & \tilde{g}[4] & \tilde{h}[2] & \tilde{g}[2] & \ddots \\ & 0 & 0 & \tilde{h}[5] & \tilde{g}[5] & \tilde{h}[3] & \tilde{g}[3] & \ddots \\ & 0 & 0 & 0 & 0 & \tilde{h}[4] & \tilde{g}[4] & \ddots \\ & 0 & 0 & 0 & 0 & \tilde{h}[5] & \tilde{g}[5] & \ddots \\ & \vdots & \vdots & \vdots & \vdots & & \ddots & \ddots \end{bmatrix} \tag{10.9}$$

for a length 6 orthogonal filter. Compared to Chap. 7 we have changed the notation, such that the synthesis filter pair is now denoted by $\tilde{\mathbf{h}}, \tilde{\mathbf{g}}$. In Chap. 7 the pair was denoted by $\mathbf{g}_0, \mathbf{g}_1$. The verification of the structure shown in (10.9) is left as an exercise.

In the same way we can write $\mathbf{T}_s$ for biorthogonal filters, except with the obvious difference that we do not have the close connection between analysis and synthesis that characterized the orthogonal filters. We will instead give an example showing how to determine $\mathbf{T}_s$ in the biorthogonal case. The biorthogonal filter pair CDF(2,4) is given by

$$\mathbf{h} = \frac{\sqrt{2}}{128} \begin{bmatrix} 3 & -6 & -16 & 38 & 90 & 38 & -16 & -6 & 3 \end{bmatrix},$$

$$\mathbf{g} = \frac{\sqrt{2}}{4} \begin{bmatrix} 1 & -2 & 1 \end{bmatrix},$$

and from (7.35) and (7.36) it follows that

$$\tilde{\mathbf{h}} = \frac{\sqrt{2}}{4} \begin{bmatrix} 1 & 2 & 1 \end{bmatrix},$$

$$\tilde{\mathbf{g}} = \frac{\sqrt{2}}{128} \begin{bmatrix} 3 & 6 & -16 & -38 & 90 & -38 & -16 & 6 & 3 \end{bmatrix}.$$

The analysis matrix becomes

$$\mathbf{T_a} = \begin{bmatrix} \ddots & \ddots & \ddots & \ddots & \ddots & \ddots & \ddots \\ \ddots & h[6] & h[5] & h[4] & h[3] & h[2] & h[1] & h[0] & 0 & 0 & 0 & 0 & 0 & \cdots \\ \ddots & 0 & 0 & g[2] & g[1] & g[0] & 0 & 0 & 0 & 0 & 0 & 0 & 0 & \cdots \\ & h[8] & h[7] & h[6] & h[5] & h[4] & h[3] & h[2] & h[1] & h[0] & 0 & 0 & 0 & \cdots \\ \cdots & 0 & 0 & 0 & 0 & g[2] & g[1] & g[0] & 0 & 0 & 0 & 0 & 0 & \cdots \\ \cdots & 0 & 0 & h[8] & h[7] & h[6] & h[5] & h[4] & h[3] & h[2] & h[1] & h[0] & 0 & \cdots \\ \cdots & 0 & 0 & 0 & 0 & 0 & 0 & g[2] & g[1] & g[0] & 0 & 0 & 0 \\ \cdots & 0 & 0 & 0 & 0 & h[8] & h[7] & h[6] & h[5] & h[4] & h[3] & h[2] & h[1] & \ddots \\ \cdots & 0 & 0 & 0 & 0 & 0 & 0 & 0 & 0 & g[2] & g[1] & g[0] & 0 & \ddots \\ & & & & & & \ddots & \ddots & \ddots & \ddots & \ddots & \ddots \end{bmatrix}$$

and, just as it was the case for orthogonal filters, the synthesis matrix consists of the synthesis IR in forward order in the columns of $\mathbf{T_s}$, such that

$$\mathbf{T_s} = \mathbf{T_a}^{-1} = \begin{bmatrix} \ddots & \ddots & \ddots & \ddots \\ \ddots & \tilde{h}[0] & \tilde{g}[2] & 0 & \tilde{g}[0] & 0 & 0 & 0 & 0 & 0 & 0 & \cdots \\ \ddots & \tilde{h}[1] & \tilde{g}[3] & 0 & \tilde{g}[1] & 0 & 0 & 0 & 0 & 0 & 0 & \cdots \\ \ddots & \tilde{h}[2] & \tilde{g}[4] & \tilde{h}[0] & \tilde{g}[2] & 0 & \tilde{g}[0] & 0 & 0 & 0 & 0 & \cdots \\ \ddots & 0 & \tilde{g}[5] & \tilde{h}[1] & \tilde{g}[3] & 0 & \tilde{g}[1] & 0 & 0 & 0 & 0 & \cdots \\ \ddots & 0 & \tilde{g}[6] & \tilde{h}[2] & \tilde{g}[4] & \tilde{h}[0] & \tilde{g}[2] & 0 & \tilde{g}[0] & 0 & 0 & \cdots \\ \ddots & 0 & \tilde{g}[7] & 0 & \tilde{g}[5] & \tilde{h}[1] & \tilde{g}[3] & 0 & \tilde{g}[1] & 0 & 0 & \cdots \\ \cdots & 0 & \tilde{g}[8] & 0 & \tilde{g}[6] & \tilde{h}[2] & \tilde{g}[4] & \tilde{h}[0] & \tilde{g}[2] & 0 & \tilde{g}[0] \\ \cdots & 0 & 0 & 0 & \tilde{g}[7] & 0 & \tilde{g}[5] & \tilde{h}[1] & \tilde{g}[3] & 0 & \tilde{g}[1] & \ddots \\ \cdots & 0 & 0 & 0 & \tilde{g}[8] & 0 & \tilde{g}[6] & \tilde{h}[2] & \tilde{g}[4] & \tilde{h}[0] & \tilde{g}[2] & \ddots \\ \cdots & 0 & 0 & 0 & 0 & 0 & \tilde{g}[7] & 0 & \tilde{g}[5] & \tilde{h}[1] & \tilde{g}[3] & \ddots \\ \cdots & 0 & 0 & 0 & 0 & 0 & \tilde{g}[8] & 0 & \tilde{g}[6] & \tilde{h}[2] & \tilde{g}[4] & \ddots \\ & & & & & \ddots & \ddots & \ddots & \ddots & \ddots \end{bmatrix}.$$

Note that the alignment of $\tilde{\mathbf{h}}$ and $\tilde{\mathbf{g}}$ must match the alignment of $\mathbf{h}$ and $\mathbf{g}$ in $\mathbf{T_a}$. We have now constructed two matrices, which perform the orthogonal and biorthogonal wavelet transforms, when multiplied with the signal.

We have introduced the matrices of the direct and inverse transforms in order to explain how we construct boundary corrections. Computationally both filtering and lifting are much more efficient transform implementations.

10.3 Gram-Schmidt Boundary Filters

The idea behind boundary filters is to replace the filters (or lifting steps) in each end of the signal with some new filter coefficients designed to preserve both the length of the signal and the perfect reconstruction property. This idea is depicted in Fig. 10.2. We start by looking more carefully at the problem

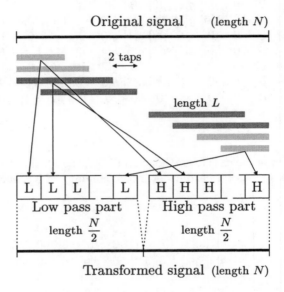

Fig. 10.2. The idea behind all types of boundary filter is to replace the filters reaching beyond the signal (see Fig. 10.1) with new, shorter filters (light grey). By having the right number of boundary filters it is possible to get exactly the same number of transform coefficients as signal samples while preserving certain properties of the wavelet transform

with zero padding. Suppose we have a finite signal **x** of length N. We first perform the zero padding, creating the new signal **s** of infinite length, by defining

$$s[n] = \begin{cases} 0 & \text{if } n \leq -1 \,, \\ x[n] & \text{if } n = 0, 1, \ldots, N-1 \,, \\ 0 & \text{if } n \geq N \,. \end{cases} \tag{10.10}$$

Suppose that the filter has length L, with the nonzero coefficients having indices between 0 and $L - 1$. To avoid special cases we also assume that N is substantially larger than L, and that both L and N are even. We then examine the formula (see (7.51))

$$(Hs)[n] = \sum_{k \in \mathbf{Z}} h[2n - k]s[k] = \sum_{k=0}^{N-1} h[2n - k]x[k]$$

for each possible value of n. If $n < 0$, the sum is always zero. The first nonzero term can occur when $n = 0$, and we have $(Hs)[0] = h[0]x[0]$. The last nonzero term occurs for $n = (N+L-2)/2$, and it is $(Hs)[(N+L-2)/2] = h[L-1]x[N-1]$. The same computation is valid for the Gs vector. Thus in the transformed signal the total number of nonzero terms can be up to $N+L-2$.

This computation also shows that in the index range $L/2 < n < N-(L/2)$ all filter coefficients are multiplied with x-entries. At the start and the end only some filter coefficients are needed, the others being multiplied by zero from the zero padding of the signal s. This leads to the introduction of the boundary filters. We modify the filters during the $L/2$ evaluations at both the beginning and the end of the signal, taking into account only those filter coefficients that are actually needed. Thus to adjust the h filter a total of L new filters will be needed. The same number of modifications will be needed for the high pass filter. It turns out that we can manage with fewer modified filters, if we shift the location of the finite signal one unit.

Let us repeat the computation above with the following modification of the zero padding. We define

$$s_{\text{shift}}[n] = \begin{cases} 0 & \text{if } n \le -2 \,, \\ x[n+1] & \text{if } n = -1, 0, 1, \ldots, N-2 \,, \\ 0 & \text{if } n \ge N-1 \,. \end{cases} \tag{10.11}$$

With this modification we find that the first nonzero term in Hs can be

$$(Hs_{\text{shift}})[0] = h[1]x[0] + h[0]x[1] \,,$$

and the last nonzero term can be

$$(Hs_{\text{shift}})[(N+L)/2 - 2] = h[L-1]x[N-2] + h[L-2]x[N-1] \,,$$

due to the assumption that L is even. With this shift we need a total of $L-2$ corrections at each end. We will use this shifted placement of the nonzero coefficients in the next subsection.

10.3.1 The DWT Matrix Applied to Finite Signals

Instead of using zero padding we could truncate the matrices $\mathbf{T_a}$ and $\mathbf{T_s}$, by removing the parts multiplying the zero padded parts of the signal. Although this gives finite matrices it does not solve the problem that the transformed signal can have more nonzero entries than the original signal. The next step is therefore to alter the truncated matrices to get orthogonal matrices. We treat only orthogonal filters, since the biorthogonal case is rather complicated.

Let us start with the example from the previous section. For a filter of length 6 and a signal of length 8 the transformed signal can have 12 non-vanishing elements, as was shown above. Let us remove the part of the matrix that multiplies zeroes in s_{shift}. The reduced matrix is denoted by T'_a, and it is given as

$$T'_a x = \begin{bmatrix} h[1] & h[0] & 0 & 0 & 0 & 0 & 0 & 0 \\ g[1] & g[0] & 0 & 0 & 0 & 0 & 0 & 0 \\ h[3] & h[2] & h[1] & h[0] & 0 & 0 & 0 & 0 \\ g[3] & g[2] & g[1] & g[0] & 0 & 0 & 0 & 0 \\ h[5] & h[4] & h[3] & h[2] & h[1] & h[0] & 0 & 0 \\ g[5] & g[4] & g[3] & g[2] & g[1] & g[0] & 0 & 0 \\ 0 & 0 & h[5] & h[4] & h[3] & h[2] & h[1] & h[0] \\ 0 & 0 & g[5] & g[4] & g[3] & g[2] & g[1] & g[0] \\ 0 & 0 & 0 & 0 & h[5] & h[4] & h[3] & h[2] \\ 0 & 0 & 0 & 0 & g[5] & g[4] & g[3] & g[2] \\ 0 & 0 & 0 & 0 & 0 & 0 & h[5] & h[4] \\ 0 & 0 & 0 & 0 & 0 & 0 & g[5] & g[4] \end{bmatrix} \begin{bmatrix} x[0] \\ x[1] \\ x[2] \\ x[3] \\ x[4] \\ x[5] \\ x[6] \\ x[7] \end{bmatrix} = \begin{bmatrix} y[0] \\ y[1] \\ y[2] \\ y[3] \\ y[4] \\ y[5] \\ y[6] \\ y[7] \\ y[8] \\ y[9] \\ y[10] \\ y[11] \end{bmatrix}. \quad (10.12)$$

It is evident from the two computations above with the original and the shifted signal that the truncation of the T_a matrix is not unique. As described above, we have chosen to align the first non-vanishing element in x with $h[1]$ and $g[1]$. This makes T'_a "more symmetric" than if we had chosen $h[0]$ and $g[0]$. Moreover, choosing the symmetric truncation guarantees linear independence of the rows, see [11], a property which we will need later. By truncating T_a to make the 12×8-matrix T'_a we have not erased any information in the transformed signal. Hence it is possible, by reducing T_s to an 8×12 matrix, to reconstruct the original signal (see Exer. 10.6).

Now we want to change the matrix T'_a such that y has the same number of coefficients as x. When looking at the matrix equation (10.12) the first idea might be to further reduce the size of T'_a, this time making an 8×8 matrix, by removing the two upper and lower most rows. The resulting matrix is denoted by T''_a. At least this will ensure a transformed signal with only 8 coefficients. By removing the two first and two last columns in T'_s we get an 8×8 synthesis matrix. The question is now whether we can reconstruct x from y or not. As before, if we can prove $T''_s T''_a = I$, perfect reconstruction is guaranteed. Although it is easily shown that we cannot obtain this (see Exer. 10.7), the truncation procedure is still useful. For it turns out that the matrices T''_s and T''_a have a very nice property which, assisted by a slight adjustment of the matrices, will lead to perfect reconstruction. Moreover this adjustment also ensures energy preservation, which is one of the properties of orthogonal filters that we want to preserve in the modified matrix.

We start by examining the truncated matrix $\mathbf{T}_a''$, which we now denote by $\mathbf{M}$, rewritten to consist of 8 row vectors

$$
\mathbf{T}_a'' = \begin{bmatrix} h[3] & h[2] & h[1] & h[0] & 0 & 0 & 0 & 0 \\ g[3] & g[2] & g[1] & g[0] & 0 & 0 & 0 & 0 \\ h[5] & h[4] & h[3] & h[2] & h[1] & h[0] & 0 & 0 \\ g[5] & g[4] & g[3] & g[2] & g[1] & g[0] & 0 & 0 \\ 0 & 0 & h[5] & h[4] & h[3] & h[2] & h[1] & h[0] \\ 0 & 0 & g[5] & g[4] & g[3] & g[2] & g[1] & g[0] \\ 0 & 0 & 0 & 0 & h[5] & h[4] & h[3] & h[2] \\ 0 & 0 & 0 & 0 & g[5] & g[4] & g[3] & g[2] \end{bmatrix} = \mathbf{M} = \begin{bmatrix} \mathbf{m}_0 \\ \mathbf{m}_1 \\ \mathbf{m}_2 \\ \mathbf{m}_3 \\ \mathbf{m}_4 \\ \mathbf{m}_5 \\ \mathbf{m}_6 \\ \mathbf{m}_7 \end{bmatrix}, \quad (10.13)
$$

where $\mathbf{m}_n$ is the n'th row vector (note that the $\mathbf{m}_k$ vectors throughout this section are row vectors). As a consequence of (7.62), (7.66), and (7.71) most of the vectors in $\mathbf{M}$ are mutually orthogonal. Moreover, the eight vectors are linearly independent (see Exer. 10.4 for this particular matrix, and the paper [11] for a general proof). This means that all the rows can be made mutually orthogonal by the Gram-Schmidt orthogonalization procedure, which in turn means that we can transform $\mathbf{M}$ to get an orthogonal matrix. With an orthogonal matrix we can find the inverse as the transpose. Thus we have also found the synthesis matrix.

Let us recall the Gram-Schmidt orthogonalization procedure. We first recall that the inner product of two row vectors $\mathbf{u}$ and $\mathbf{v}$ can be written as the matrix product $\mathbf{uv}^\mathsf{T}$. Given a set of mutually orthogonal row vectors $\mathbf{u}_0, \dots, \mathbf{u}_N$, and a row vector $\mathbf{v}$ we get a vector orthogonal to the $\mathbf{u}_n$ by taking

$$
\mathbf{v}' = \mathbf{v} - \sum_{n=0}^{N} \frac{\mathbf{u}_n \mathbf{v}^\mathsf{T}}{\|\mathbf{u}_n\|^2} \mathbf{u}_n . \quad (10.14)
$$

If $\mathbf{v}$ is in the subspace spanned by the $\mathbf{u}$ vectors, $\mathbf{v}'$ will be the zero vector. It is easy to verify that $\mathbf{v}'$ is orthogonal to all the vectors $\mathbf{u}_n$, $n = 0, \dots, N$. Thus the set of vectors $\mathbf{u}_0, \mathbf{u}_1, \dots, \mathbf{u}_N, \mathbf{v}'$ consists of $N+2$ mutually orthogonal vectors. In this manner any set of linearly independent vectors can be transformed to a set of mutually orthogonal vectors, which span the same subspace as the original set. This is the Gram-Schmidt orthogonalization procedure. It desired, the new vectors can be normalized to have norm one, to get an orthonormal set.

We want all the rows in $\mathbf{M}$ to be mutually orthogonal. Since $\mathbf{m}_2$ through $\mathbf{m}_5$ already are orthogonal (they have not been truncated), we need only orthogonalize $\mathbf{m}_0$, $\mathbf{m}_1$, $\mathbf{m}_6$, and $\mathbf{m}_7$ with respect to the remaining vectors. We start by orthogonalizing $\mathbf{m}_0$ with respect to $\mathbf{m}_2$ through $\mathbf{m}_5$,

$$
\mathbf{m}_0' = \mathbf{m}_0 - \sum_{n=2}^{5} \frac{\mathbf{m}_n \mathbf{m}_0^\mathsf{T}}{\|\mathbf{m}_n\|^2} \mathbf{m}_n , \quad (10.15)
$$

followed by orthogonalization of $\mathbf{m}_1$ with respect to $\mathbf{m}_0'$ and $\mathbf{m}_2$ through $\mathbf{m}_5$.

$$\mathbf{m}_1' = \mathbf{m}_1 - \frac{\mathbf{m}_0'\mathbf{m}_1^\top}{\|\mathbf{m}_0'\|^2}\mathbf{m}_0' - \sum_{n=2}^{5} \frac{\mathbf{m}_n\mathbf{m}_1^\top}{\|\mathbf{m}_n\|^2}\mathbf{m}_n \ . \tag{10.16}$$

We continue with $\mathbf{m}_7$ and $\mathbf{m}_6$. Note that they are orthogonal to $\mathbf{m}_0$ and $\mathbf{m}_1$, since the nonzero entries do not overlap. Thus if we compute

$$\mathbf{m}_7' = \mathbf{m}_7 - \sum_{n=2}^{5} \frac{\mathbf{m}_n\mathbf{m}_7^\top}{\|\mathbf{m}_n\|^2}\mathbf{m}_n \ , \tag{10.17}$$

$$\mathbf{m}_6' = \mathbf{m}_6 - \frac{\mathbf{m}_7'\mathbf{m}_6^\top}{\|\mathbf{m}_7'\|^2}\mathbf{m}_7' - \sum_{n=2}^{5} \frac{\mathbf{m}_n\mathbf{m}_6^\top}{\|\mathbf{m}_n\|^2}\mathbf{m}_n \ . \tag{10.18}$$

then these vectors are also orthogonal to $\mathbf{m}_0'$ and $\mathbf{m}_1'$. Actually the number of computations can be reduced, see Exer. 10.8.

We now replace the first two rows in $\mathbf{M}$ with $\mathbf{m}_0'$ and $\mathbf{m}_1'$, respectively, and similarly with the last two rows. The rows in the new matrix are orthogonal. We then normalize them to get a new orthogonal matrix. Since $\mathbf{m}_2$ through $\mathbf{m}_5$ already have norm 1, we need only normalize the four new vectors.

$$\mathbf{m}_n'' = \frac{\mathbf{m}_n'}{\|\mathbf{m}_n'\|}, \quad n = 0, 1, 6, 7 \ .$$

The result is that we have transformed the matrix $\mathbf{M}$ to the orthogonal matrix

$$\mathbf{M}' = \begin{bmatrix} \mathbf{m}_0'' \\ \mathbf{m}_1'' \\ \mathbf{m}_2 \\ \mathbf{m}_3 \\ \mathbf{m}_4 \\ \mathbf{m}_5 \\ \mathbf{m}_6'' \\ \mathbf{m}_7'' \end{bmatrix} \ .$$

The new synthesis matrix is obtained as the transposed matrix

$$(\mathbf{M}')^{-1} = (\mathbf{M}')^\top = \left[(\mathbf{m}_0'')^\top \ (\mathbf{m}_1'')^\top \ \mathbf{m}_2^\top \ \mathbf{m}_3^\top \ \mathbf{m}_4^\top \ \mathbf{m}_5^\top \ (\mathbf{m}_6'')^\top \ (\mathbf{m}_7'')^\top \right] \ .$$

Note that the changes in the analysis matrix are only performed at the first two and last two rows. The two new top rows are called the *left boundary filters* and those at the bottom the *right boundary filters*.

If we need to transform a longer finite signal of even length, we can just add the necessary pairs of $\mathbf{h}$ and $\mathbf{g}$ in the middle, since these rows are orthogonal to the four new vectors at the top and bottom. Let us verify this claim. Let us look at for example $\mathbf{m}_1'$ from (10.16). The vectors $\mathbf{m}_0$ and $\mathbf{m}_1$ are orthogonal to the new rows in the middle, since they have no nonzero entries overlapping with the entries in the new middle rows, see (10.13). The remaining vectors in the sums defining $\mathbf{m}_1'$ are combinations of the vectors $\mathbf{m}_2, \ldots, \mathbf{m}_5$, which have not been truncated, and therefore are orthogonal to the new middle rows. The orthogonality of the three remaining vectors to the new middle rows is obtained by similar arguments.

10.3.2 The General Case

The derivation of boundary filters for a length 6 IR makes it easy to generalize the method. For any wavelet filter it is always possible to truncate the corresponding analysis matrix $\mathbf{T_a}$, such that the result is an $N \times N$ matrix $\mathbf{M}$ (with N even) with all but the first and last $L/2 - 1$ rows containing whole IRs, and such that the upper and lower truncated rows have an equal number non-vanishing entries. If $L = 4K + 2$, $K \in \mathbf{N}$, the first row in $\mathbf{M}$ will be (a part of) the low pass IR $\mathbf{h}$, and if $L = 4K$ the first row will be (a part of) the high pass IR $\mathbf{g}$, see Exer. 10.5 It can be shown (see [10]) that this symmetric truncation always produces a full rank matrix (the rows are linearly independent). As described above this guarantees that we can apply the Gram-Schmidt orthogonalization procedure to get a new orthogonal matrix. The truncation of the infinite transform matrix with a filter of length L is thus of the form

$$\mathbf{M} = \begin{bmatrix} \mathbf{m}_0 \\ \vdots \\ \mathbf{m}_{L/2-2} \\ \mathbf{m}_{L/2-1} \\ \vdots \\ \mathbf{m}_{N-L/2+2} \\ \mathbf{m}_{N-L/2+1} \\ \vdots \\ \mathbf{m}_{N-1} \end{bmatrix} \left. \begin{matrix} \\ \\ \\ \end{matrix} \right\} L/2 - 1 \text{ left truncated IRs },$$
$$\left. \begin{matrix} \\ \\ \end{matrix} \right\} N - L + 2 \text{ whole IRs }, \qquad (10.19)$$
$$\left. \begin{matrix} \\ \\ \\ \end{matrix} \right\} L/2 - 1 \text{ right truncated IRs }.$$

Then all the truncated rows are orthogonalized by the Gram-Schmidt procedure (10.14). It is easy to show that (see Exer. 10.8) we need only orthogonalize $\mathbf{m}_0$ through $\mathbf{m}_{L/2-2}$ with respect to themselves (and not to all the IRs). So the left boundary filters $\mathbf{m}_k^l$ are defined as

$$\mathbf{m}_k' = \mathbf{m}_k - \sum_{n=0}^{k-1} \frac{\mathbf{m}_n \mathbf{m}_k^\top}{\|\mathbf{m}_n\|^2} \mathbf{m}_n , \qquad k = 0, 1, \ldots, L/2 - 2 ,$$

and

$$\mathbf{m}_k^l = \frac{\mathbf{m}_k'}{\|\mathbf{m}_k'\|^2} \, , \qquad k = 0, \dots, L/2 - 2 \, .$$

In the same way the vectors $\mathbf{m}_{N-L/2+2}$ through $\mathbf{m}_{N-1}$ are converted into $L/2-1$ right boundary filters, which we denote by $\mathbf{m}_0^r$ through $\mathbf{m}_{L/2-1}^r$. The Gram-Schmidt orthogonalization of the right boundary filters starts with $\mathbf{m}_{N-1}$. The new orthogonal matrix then becomes

$$\mathbf{M}' = \begin{bmatrix} \mathbf{m}_0^l \\ \vdots \\ \mathbf{m}_{L/2-2}^l \\ \mathbf{m}_{L/2-1} \\ \vdots \\ \mathbf{m}_{N-L/2+2} \\ \mathbf{m}_0^r \\ \vdots \\ \mathbf{m}_{L/2-2}^r \end{bmatrix} \begin{matrix} \left.\vphantom{\begin{matrix}a\\b\\c\end{matrix}}\right\} \; L/2-1 \text{ left boundary filters} , \\ \\ \left.\vphantom{\begin{matrix}a\\b\\c\end{matrix}}\right\} \; N-L+2 \text{ whole filters} , \\ \\ \left.\vphantom{\begin{matrix}a\\b\\c\end{matrix}}\right\} \; L/2-1 \text{ right boundary filters} , \end{matrix} \qquad (10.20)$$

The length of the boundary filter $\mathbf{m}_{L/2-2}^l$ constructed this way is $L-2$, and the length of $\mathbf{m}_k^l$ is decreasing with k. The right boundary filters exhibit the same structure.

The boundary filters belonging to the inverse transform are easily found, since the synthesis matrix is the transpose of analysis matrix. The implementation of the construction and the use of the boundary filters are both demonstrated in Chap. 11.

10.4 Periodization

The simple solution to the boundary problem was zero padding. Another possibility is to choose samples from the signal to use for the missing samples. One way of doing this is to *periodize* the finite signal. Suppose the original finite signal is the column vector $\mathbf{x}$, of length N. Then the periodized signal is given as

$$\mathbf{x}^p = \begin{bmatrix} \vdots \\ \mathbf{x} \\ \mathbf{x} \\ \mathbf{x} \\ \vdots \end{bmatrix} = \begin{bmatrix} \cdots & x[N-2] & x[N-1] & x[0] & \cdots & x[N-1] & x[0] & x[1] & \cdots \end{bmatrix}^\top .$$

This signal is periodic with period N, since $x^p[k+N] = x^p[k]$ for all integers k. It is important to note that the signal $\mathbf{x}^p$ has infinite energy. But we can still transform it with $\mathbf{T_a}$, since we use filters of finite length, such that each row in $\mathbf{T_a}$ only has a finite number of nonzero entries. Let $\mathbf{y^p} = \mathbf{T_a}\mathbf{x^p}$, or explicitly

$$\mathbf{y^p} = \mathbf{T_a}\mathbf{x^p} =$$

$$
\begin{bmatrix}
\cdot & \cdot & \cdot & \cdot & \cdot & \cdot & & & & \\
h[5] & h[4] & h[3] & h[2] & h[1] & h[0] & 0 & 0 & 0 & 0 & \cdot \\
g[5] & g[4] & g[3] & g[2] & g[1] & g[0] & 0 & 0 & 0 & 0 & \cdot \\
\cdot & 0 & 0 & h[5] & h[4] & h[3] & h[2] & h[1] & h[0] & 0 & 0 & \cdot \\
\cdot & 0 & 0 & g[5] & g[4] & g[3] & g[2] & g[1] & g[0] & 0 & 0 & \cdot \\
\cdot & 0 & 0 & 0 & 0 & h[5] & h[4] & h[3] & h[2] & h[1] & h[0] \\
\cdot & 0 & 0 & 0 & 0 & g[5] & g[4] & g[3] & g[2] & g[1] & g[0] \\
& & & & \cdot & \cdot & \cdot & \cdot & \cdot & \cdot & \cdot \\
\end{bmatrix}
\begin{bmatrix}
\vdots \\
x[N-2] \\
x[N-1] \\
x[0] \\
\vdots \\
x[N-1] \\
x[0] \\
x[1] \\
\vdots
\end{bmatrix}
=
\begin{bmatrix}
\vdots \\
y[N-2] \\
y[N-1] \\
y[0] \\
\vdots \\
y[N-1] \\
y[0] \\
y[1] \\
\vdots
\end{bmatrix}.
$$

$$(10.21)$$

The transformed signal is also periodic with period N. We leave the easy verification as Exer. 10.9. We select N consecutive entries in $\mathbf{y^p}$ to represent it. The choice of these entries is not unique, but below we see that a particular choice is preferable, to match up with the given signal $\mathbf{x}$.

We have transformed a finite signal $\mathbf{x}$ into another finite signal $\mathbf{y}$ of equal length. The same procedure can be used to inversely transform $\mathbf{y}$ into $\mathbf{x}$ using the infinite $\mathbf{T_s}$ (see Exer. 10.9). Thus periodization is a way of transforming a finite signal while preserving the length of it. In implementations we need to use samples from $\mathbf{x}$ instead of the zero samples used in zero padding. We only need enough samples to cover the extent of the filters, which is at most $L-2$. But we would like to avoid extending the signal at all, since this requires extra time and memory in an implementation. Fortunately it is very easy to alter the transform matrix to accommodate this desire. This means that we can transform $\mathbf{x}$ directly into $\mathbf{y}$.

We start by reducing the infinite transform matrix such that it fits the signal. If the signal has length N, we reduce the matrix to an $N \times N$ matrix, just at we did in the previous section on boundary filters. Although we do not need a symmetric structure of the matrix this time, we choose symmetry anyway in order to obtain a transformed signal of the same form as the original one.

Let us use the same example as above. For a signal of length 10 and filter of length 6 the reduced matrix is

$$\begin{bmatrix}
h[3] & h[2] & h[1] & h[0] & 0 & 0 & 0 & 0 & 0 & 0 \\
g[3] & g[2] & g[1] & g[0] & 0 & 0 & 0 & 0 & 0 & 0 \\
h[5] & h[4] & h[3] & h[2] & h[1] & h[0] & 0 & 0 & 0 & 0 \\
g[5] & g[4] & g[3] & g[2] & g[1] & g[0] & 0 & 0 & 0 & 0 \\
0 & 0 & h[5] & h[4] & h[3] & h[2] & h[1] & h[0] & 0 & 0 \\
0 & 0 & g[5] & g[4] & g[3] & g[2] & g[1] & g[0] & 0 & 0 \\
0 & 0 & 0 & 0 & h[5] & h[4] & h[3] & h[2] & h[1] & h[0] \\
0 & 0 & 0 & 0 & g[5] & g[4] & g[3] & g[2] & g[1] & g[0] \\
0 & 0 & 0 & 0 & 0 & 0 & h[5] & h[4] & h[3] & h[2] \\
0 & 0 & 0 & 0 & 0 & 0 & g[5] & g[4] & g[3] & g[2]
\end{bmatrix}.$$

The periodization is accomplished by inserting all the deleted filter coefficients in appropriate places in the matrix. This changes the matrix to

$$\mathbf{T}_a^p = \begin{bmatrix}
h[3] & h[2] & h[1] & h[0] & 0 & 0 & 0 & 0 & h[5] & h[4] \\
g[3] & g[2] & g[1] & g[0] & 0 & 0 & 0 & 0 & g[5] & g[4] \\
h[5] & h[4] & h[3] & h[2] & h[1] & h[0] & 0 & 0 & 0 & 0 \\
g[5] & g[4] & g[3] & g[2] & g[1] & g[0] & 0 & 0 & 0 & 0 \\
0 & 0 & h[5] & h[4] & h[3] & h[2] & h[1] & h[0] & 0 & 0 \\
0 & 0 & g[5] & g[4] & g[3] & g[2] & g[1] & g[0] & 0 & 0 \\
0 & 0 & 0 & 0 & h[5] & h[4] & h[3] & h[2] & h[1] & h[0] \\
0 & 0 & 0 & 0 & g[5] & g[4] & g[3] & g[2] & g[1] & g[0] \\
h[1] & h[0] & 0 & 0 & 0 & 0 & h[5] & h[4] & h[3] & h[2] \\
g[1] & g[0] & 0 & 0 & 0 & 0 & g[5] & g[4] & g[3] & g[2]
\end{bmatrix}. \qquad (10.22)$$

It can be shown (see Exer. 10.9) that $\mathbf{y} = \mathbf{T}_a^p \mathbf{x}$ is the same signal as found in (10.21). Now $\mathbf{T}_a^p$ is orthogonal, so the inverse transform is given by $(\mathbf{T}_a^p)^\top$.

The same principle can be applied to biorthogonal filters. A length 12 signal and a biorthogonal filter set with analysis low pass filter of length 9 and high pass filter of length 3 would give rise to the matrix

$$\mathbf{T}_a^p = \begin{bmatrix}
h[5] & h[4] & h[3] & h[2] & h[1] & h[0] & 0 & 0 & 0 & h[8] & h[7] & h[6] \\
0 & g[2] & g[1] & g[0] & 0 & 0 & 0 & 0 & 0 & 0 & 0 & 0 \\
h[7] & h[6] & h[5] & h[4] & h[3] & h[2] & h[1] & h[0] & 0 & 0 & 0 & h[8] \\
0 & 0 & 0 & g[2] & g[1] & g[0] & 0 & 0 & 0 & 0 & 0 & 0 \\
0 & h[8] & h[7] & h[6] & h[5] & h[4] & h[3] & h[2] & h[1] & h[0] & 0 & 0 \\
0 & 0 & 0 & 0 & 0 & g[2] & g[1] & g[0] & 0 & 0 & 0 & 0 \\
0 & 0 & 0 & h[8] & h[7] & h[6] & h[5] & h[4] & h[3] & h[2] & h[1] & h[0] \\
0 & 0 & 0 & 0 & 0 & 0 & 0 & g[2] & g[1] & g[0] & 0 & 0 \\
h[1] & h[0] & 0 & 0 & 0 & h[8] & h[7] & h[6] & h[5] & h[4] & h[3] & h[2] \\
0 & 0 & 0 & 0 & 0 & 0 & 0 & 0 & 0 & g[2] & g[1] & g[0] \\
h[3] & h[2] & h[1] & h[0] & 0 & 0 & 0 & h[8] & h[7] & h[6] & h[5] & h[4] \\
g[1] & g[0] & 0 & 0 & 0 & 0 & 0 & 0 & 0 & 0 & 0 & g[2]
\end{bmatrix}. \qquad (10.23)$$

The corresponding synthesis matrix cannot be found simply by transposing the analysis matrix, since it is not orthogonal. It is easily constructed, however.

$$
\mathbf{T_s^p} =
\begin{bmatrix}
\tilde{h}[0] & \tilde{g}[2] & 0 & \tilde{g}[0] & 0 & 0 & 0 & \tilde{g}[8] & 0 & \tilde{g}[6] & \tilde{h}[2] & \tilde{g}[4] \\
\tilde{h}[1] & \tilde{g}[3] & 0 & \tilde{g}[1] & 0 & 0 & 0 & 0 & 0 & \tilde{g}[7] & 0 & \tilde{g}[5] \\
\tilde{h}[2] & \tilde{g}[4] & \tilde{h}[0] & \tilde{g}[2] & 0 & \tilde{g}[0] & 0 & 0 & 0 & \tilde{g}[8] & 0 & \tilde{g}[6] \\
0 & \tilde{g}[5] & \tilde{h}[1] & \tilde{g}[3] & 0 & \tilde{g}[1] & 0 & 0 & 0 & 0 & 0 & \tilde{g}[7] \\
0 & \tilde{g}[6] & \tilde{h}[2] & \tilde{g}[4] & \tilde{h}[0] & \tilde{g}[2] & 0 & \tilde{g}[0] & 0 & 0 & 0 & \tilde{g}[8] \\
0 & \tilde{g}[7] & 0 & \tilde{g}[5] & \tilde{h}[1] & \tilde{g}[3] & 0 & \tilde{g}[1] & 0 & 0 & 0 & 0 \\
0 & \tilde{g}[8] & 0 & \tilde{g}[6] & \tilde{h}[2] & \tilde{g}[4] & \tilde{h}[0] & \tilde{g}[2] & 0 & \tilde{g}[0] & 0 & 0 \\
0 & 0 & 0 & \tilde{g}[7] & 0 & \tilde{g}[5] & \tilde{h}[1] & \tilde{g}[3] & 0 & \tilde{g}[1] & 0 & 0 \\
0 & 0 & 0 & \tilde{g}[8] & 0 & \tilde{g}[6] & \tilde{h}[2] & \tilde{g}[4] & \tilde{h}[0] & \tilde{g}[2] & 0 & \tilde{g}[0] \\
0 & 0 & 0 & 0 & 0 & \tilde{g}[7] & 0 & \tilde{g}[5] & \tilde{h}[1] & \tilde{g}[3] & 0 & \tilde{g}[1] \\
0 & \tilde{g}[0] & 0 & 0 & 0 & \tilde{g}[8] & 0 & \tilde{g}[6] & \tilde{h}[2] & \tilde{g}[4] & \tilde{h}[0] & \tilde{g}[2] \\
0 & \tilde{g}[1] & 0 & 0 & 0 & 0 & 0 & \tilde{g}[7] & 0 & \tilde{g}[5] & \tilde{h}[1] & \tilde{g}[3]
\end{bmatrix}. \tag{10.24}
$$

As before, perfect reconstruction is guaranteed, since $\mathbf{T_s^p T_a^p} = \mathbf{I}$. This can be made plausible by calculating the matrix product of the first row in $\mathbf{T_s^p}$ and the first column in $\mathbf{T_a^p}$, which gives

$$
\tilde{h}[0]h[5] + \tilde{h}[2]h[3] + \tilde{g}[4]g[1] . \tag{10.25}
$$

To further process this formula, we need the relations between g, $\tilde{g}$ and h, $\tilde{h}$. They are given by (7.35) and (7.36) on p. 71. First we need to determine k and c. If we let the z-transform of h and g in (10.23) be given by

$$
H(z) = \sum_{n=0}^{8} h[n]z^{-n} \quad \text{and} \quad G(z) = \sum_{n=0}^{2} g[n]z^{-n} ,
$$

and calculate the determinant (7.34), we find that (remember that a different notation for filters is used in Chap. 7)

$$
H(z)G(-z) - G(z)H(-z) = -2z^{-5} ,
$$

and hence $k = 2$ and $c = -1$. From (7.35) it then follows that

$$
\tilde{H}(z) = -z^5 G(-z) . \tag{10.26}
$$

The odd shift in index due to the power z^5 is a consequence of the perfect reconstruction requirement, as explained in Chap. 7. The immediate result of (10.26) is

$$
\sum_{n} \tilde{h}[n]z^{-n} = -\sum_{n}(-1)^n g[n]z^{-n+5} ,
$$

and if we assume that $g[n] \neq 0$ for $n = 0, 1, 2$ (as we implicitly did in (10.23)), then we find that

$$
\tilde{h}[-5]z^5 + \tilde{h}[-4]z^4 + \tilde{h}[-3]z^3 = -g[0]z^5 + g[1]z^4 - g[2]z^3 , \tag{10.27}
$$

which seems to match poorly with our choice of index of $\tilde{h}[n]$ in (10.24). The reason is that while the index $n = -5, -4, -3$ of $\tilde{h}[n]$ is the correct one in the sense that it matches in the z-transform, we usually choose a more convenient indexing in implementations (like $n = 0, 1, 2$). To complete the calculation started in (10.25), we need to stick to the correct indexing, however. We therefore substitute $g[1] = \tilde{h}[-4]$. Doing the same calculation for $\tilde{g}[n]$, we find the we can substitute $\tilde{g}[4] = h[4]$ (this is left as an exercise). Now (10.25), with the correct indexing of $\tilde{h}$, becomes

$$\tilde{h}[-5]h[5] + \tilde{h}[-3]h[3] + \tilde{h}[-4]h[4]$$
$$= \sum_{k=3}^{5} \tilde{h}[-k]h[k] = \sum_{k} \tilde{h}[-k]h[k] = 1 \ . \quad (10.28)$$

The second equality is valid since $\tilde{h}[n] = 0$ except for $n = -5, -4, -3$, and the third follows immediately from (7.56).

10.4.1 Mirroring

Finally, let us briefly describe a variant of periodization. One can take a finite signal $[x[0] \ \cdots \ x[N-1]]$, and first mirror it to get the signal $[x[0] \ \ldots \ x[N-1] \ x[N-1] \ \ldots, x[0]]$ of length $2N$. Then one can apply periodization to this signal. The above procedure can then be used to get a truncated transformation matrix, of size $2N \times 2N$. It is in general not possible to get a truncated matrix of size $N \times N$, which is orthogonal.

Let us briefly discuss the difference between periodization and mirroring. In Fig. 10.3 we have shown a continuous signal on the interval from 0 to T, which in the top part has been periodized with period T, and in the bottom part has been mirrored, to give a signal periodic with period $2T$.

Sampling these two signals leads to discrete signals that have been periodized or mirrored from the samples located between 0 and T. Two problems are evident. The periodization can lead to jump discontinuities at the points of continuation, whereas mirroring leads to discontinuities in the first derivative, unless this derivative is zero at the continuation points. These singularities then show up in the wavelet analysis of the discrete signals as large coefficients at some scale. They are artifacts produced by our boundary correction method. Mirroring is often used in connection with images, as in the separable transforms discussed in Chap. 6, since the eye is very sensitive to asymmetry.

10.5 Moment Preserving Boundary Filters

The two methods for handling the boundary problem presented so far have focused on maintaining the orthogonality of the transform. Orthogonality is

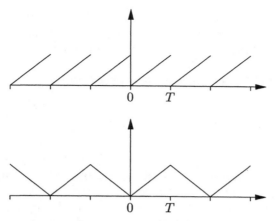

Fig. 10.3. The top part shows periodic continuation of a signal, and the bottom part mirroring

important, since it is equivalent with energy preservation. But there are other properties beyond energy that it can be useful to preserve under transformation. One of them is related to moments of a sequence.

At the end of Sect. 3.3 we introduced the term *moment* of a sequence. We derived

$$\sum_n n s_{j-1}[n] = \tfrac{1}{2} \sum_n n s_j[n] \,, \tag{10.29}$$

which shows that the transform discussed in Sect. 3.3 (the CDF(2,2) transform) preserves the *first moment* of a sequence. Generally, we say that

$$\sum_n n^k s[n] \tag{10.30}$$

is the k'th moment of the sequence **s**. Since (10.29) was derived without taking undefined samples into account, it applies to infinite sequences only. A finite sequence can be made infinite by zero padding, but this method causes the sequence s_{j-1} to be more than half the length of sequence s_j. If we want (10.29) to be valid for finite sequence of the same length, a better method is needed. The next two subsections discuss two important questions regarding the preservation of moments, namely why and how.

10.5.1 Why Moment Preserving Transforms?

To answer this question we will start by making some observations on the high pass IR **g** of a wavelet transform. For all such IRs it is true that

$$\sum_n n^k g[n] = 0, \quad k = 0, \dots, M-1, \tag{10.31}$$

for some $M \geq 1$. The M depends on the filter. For Daubechies 4 this property holds for $M = 2$, while for CDF(2,2) we have $M = 2$, and for CDF(4,6) we have $M = 4$. A sequence satisfying (10.31) for some M is said to have M *vanishing moments*.

Assume that the filter **g** has M vanishing moments. Take a polynomial

$$p(t) = \sum_{j=0}^{M-1} p_j t^j$$

of degree at most $M - 1$. We then take a signal obtained by sampling this polynomial at the integers, i.e. we take $s[n] = p(n)$. Now we filter this signal with **g**. In the following computation we first change the summation variable, then insert the polynomial expression for the signal, expand $(n - k)^j$ using the binomial formula, and finally change the order of summation.

$$
\begin{aligned}
(\mathbf{g} * \mathbf{s})[n] &= \sum_k g[n - k]s[n] \\
&= \sum_k g[k]s[n - k] \\
&= \sum_k g[k] \sum_{j=0}^{M-1} p_j(n - k)^j \\
&= \sum_k g[k] \sum_{j=0}^{M-1} p_j \sum_{m=0}^{j} \binom{j}{m}(-1)^m k^m n^{j-m} \\
&= \sum_{j=0}^{M-1} p_j \sum_{m=0}^{j} \binom{j}{m}(-1)^m n^{j-m} \sum_k k^m g[k] \\
&= 0 .
\end{aligned}
\tag{10.32}
$$

Note that we work with a filter **g** of finite length, so all sum above are finite. Thus filtering with **g** maps a signal obtained from sampling a polynomial of degree at most $M - 1$ to zero. Note also that we do not have to sample the polynomial at the integers. It is enough that the sample points are equidistant.

This property of the high pass filter **g** has an interesting consequence, when we do one step in a DWT. Since we have perfect reconstruction, the polynomial samples get mapped into the low pass part. This is consistent with the intuitive notion that polynomials of low degree do not oscillate much, meaning that they contain no high frequencies. The computation in (10.32) shows that with the particular filters used here the high pass part is actually zero, and not just close to zero, which is typical for non-ideal filters (at this point one should recall the filters used in a wavelet decomposition are not ideal).

Due to these properties it would be interesting to have a boundary correction method which preserved vanishing moments of finite signals of a given length. Such a method was found by A. Cohen, I. Daubechies, and P. Vial [3]. Their solution is presented briefly in the next section.

10.5.2 How to Make Moment Preserving Transforms

The idea for preserving the number of vanishing moments, and hence be able to reproduce polynomials completely in the low pass part, is simple, although the computations are non-trivial. We will only show the basic steps, and leave out most of the computations.

We start our considerations by redoing the computation in (10.32), this time for a general filter $\mathbf{h}$ of length L, and the same signal $\mathbf{s}$. This time we choose the opposite order in the application of the binomial formula. In the second equality we use (7.9). Otherwise the computations are identical. We get

$$
\begin{aligned}
(\mathbf{h} * \mathbf{s})[n] &= \sum_{k=n-L+1}^{n} h[n-k]s[n] \\
&= \sum_{k=0}^{L-1} h[k]s[n-k] \\
&= \sum_{k=0}^{L-1} h[k] \sum_{j=0}^{M-1} p_j(n-k)^j \\
&= \sum_{k=0}^{L-1} h[k] \sum_{j=0}^{M-1} p_j \sum_{m=0}^{j} \binom{j}{m} n^m (-k)^{j-m} \\
&= \sum_{m=0}^{M-1} q_m n^m \,,
\end{aligned}
\tag{10.33}
$$

where we have not written out the complicated expressions for the coefficients q_m, since they are not needed. We see from this computation that convolution with any filter $\mathbf{h}$ of finite length takes a sampled polynomial of degree at most $M-1$ into another sampled polynomial, again of degree at most $M-1$. If we have a finite number of nonzero samples, then the resulting convolution will have more samples, as explained above.

We would like to be able to invert this computation, in the sense that we start with the signal $\mathbf{x}$ of length N, obtained by sampling an arbitrary polynomial of degree at most $M-1$,

$$
x[n] = \sum_{m=0}^{M-1} q_m n^m, \quad n = 0, \ldots, N-1 \,,
$$

and would like to find another polynomial p of degree at most $M-1$, and a signal $\mathbf{s}$ of the same length N, obtained by sampling p, such that $\mathbf{x} = \mathbf{h} * \mathbf{s}$. To do this for all polynomials of degree at most $M-1$ is equivalent to the boundary filter constructions already done in Sect. 10.3.

We want corrections to the filters used at the start and end of the signal in order to preserve vanishing moments for signals of a fixed finite length. It is done as follows. The first (leftmost) boundary filter on the left and the last (rightmost) boundary filter on the right is chosen such that they preserve vanishing of the moment of order $m = 0$ in the high pass part. The next pair is chosen such that moments or order $m = 1$ vanish. We continue until we reach the value of M for the transform under consideration.

It is by no means trivial to construct the boundary filters and to prove that the described procedure does produce a moment preserving transform, and it takes further computations to make these new boundary filters both orthogonal and of decreasing length (as we did with the Gram-Schmidt boundary filters, see the description at the end of Sect. 10.3.2).

10.5.3 Use of the Boundary Filters

Unfortunately, the efforts so far are not enough to construct a transform applicable to finite signals, such that it preserves vanishing moments. Thus the description above was incomplete. It must remain so, since this is a very technical question, beyond the scope of this book. Briefly, what remains to be done is an extra step, which consists in pre-conditioning the signal prior to transformation by multiplying the first M and the last M samples by an $M \times M$ matrix. After transformation we multiply the result by the inverse of this matrix, at the beginning and end of the signal.

The software available electronically, see Chap. 14, contains functions implementing this procedure. See the documentation provided there for further information and explanation.

Exercises

10.1 Start by reviewing results from linear algebra on orthogonal matrices and the Gram-Schmidt orthogonalization procedure.

10.2 Determine which results in Chap. 7 are needed to show that the synthesis matrix $\mathbf{T}_s$ in (10.9) is the same as the transpose of the analysis matrix $\mathbf{T}_a$ in (10.7), or equivalently that $\mathbf{T}_a^\top \mathbf{T}_a = \mathbf{I}$.

10.3 Show that an orthogonal matrix preserves energy when multiplied with a signal. In other words, $\|\mathbf{T}\mathbf{x}\|^2 = \|\mathbf{x}\|^2$ whenever $\mathbf{T}$ is orthogonal. Remember that $\mathbf{T}^\top = \mathbf{T}^{-1}$.

10.4 Show that the first two rows of $\mathbf{T}_a''$ in (10.13) are linearly independent. Hint: Let $\mathbf{r}_1$ and $\mathbf{r}_2$ be the two rows, substitute the g's in $\mathbf{r}_2$ with h's, and show that there does not exist α such that $\alpha \mathbf{r}_1 = \mathbf{r}_2$.

10.5 Verify the following statement from Sect. 10.3.2: If $(L-2)/2$ is even, then the top row in the symmetrically truncated $\mathbf{T}_a$ contains coefficients from the filter $\mathbf{h}$ and the bottom row coefficients from the filter $\mathbf{g}$. If $(L-2)/2$ is odd, the positions of the filter coefficients are reversed.

10.6 The purpose of this exercise is to construct the truncated synthesis matrix $\mathbf{T}_s'$ to match the truncated analysis matrix $\mathbf{T}_a'$. Keep in mind the difference in notation between this chapter and Chap. 7.

1. Write the $\mathbf{T}_s'$ by reducing the matrix in (10.9) in such a way that its structure matches that of $\mathbf{T}_a'$ in (10.12).
2. Let $\mathbf{h}$ be an orthogonal low pass analysis filter, and define

$$H(z) = \sum_{n=0}^{5} h[n] z^{-n}.$$

   Determine the corresponding high pass analysis filter G using (7.67).
3. By using (7.34) and (7.72) determine k and c, and find by (7.35) and (7.36) the synthesis filters $\tilde{H}$ and $\tilde{G}$.
4. To verify that $\mathbf{T}_s'$ it is indeed the inverse of $\mathbf{T}_a'$, it is necessary to verify that $\mathbf{T}_s' \mathbf{T}_a' = \mathbf{I}$. Calculate two of the many inner products, starting with the inner product of

$$[\tilde{h}[4]\ \tilde{g}[4]\ \tilde{h}[2]\ \tilde{g}[2]\ \tilde{h}[0]\ \tilde{g}[0]] \quad \text{and} \quad [h[1]\ g[1]\ h[3]\ g[3]\ h[5]\ g[5]] \ .$$

   Use the correctly indexed filters, which are the four filters H, G, $\tilde{H}$, and $\tilde{G}$ found above.
5. Determine also the inner product of

$$[\tilde{h}[5]\ \tilde{g}[5]\ \tilde{h}[3]\ \tilde{g}[3]\ \tilde{h}[1]\ \tilde{g}[1]] \quad \text{and} \quad [h[1]\ g[1]\ h[3]\ g[3]\ h[5]\ g[5]] \ .$$

6. Describe why these calculations make it plausible that $\mathbf{T}_s'$ is the matrix which reconstructs a signal transformed with $\mathbf{T}_a'$. Remember that the condition for perfect reconstruction is $\mathbf{T}_s' \mathbf{T}_a' = \mathbf{I}$.

10.7 Show that the matrix $\mathbf{T}_a''$ in (10.13) is not orthogonal.

10.8 Most of the orthogonalization calculations in (10.15) – (10.18) are redundant. Verify the following statements.

1. (10.15) is unnecessary.
2. (10.16) can be reduced to

$$\mathbf{m}_1' = \mathbf{m}_1 - \frac{\mathbf{m}_0 \mathbf{m}_1^\mathsf{T}}{\|\mathbf{m}_0\|^2} \mathbf{m}_0 \ . \tag{10.34}$$

3. This result also applies to (10.17) and (10.18).
4. For any transform matrix (for any orthogonal filter of length L) we need only orthogonalize the $L/2 - 2$ upper and lower most rows.
5. The low pass left boundary filters constructed this way has staggered length, i.e. no two left boundary filters has the same number of non-vanishing filter taps.
6. The difference in number of non-vanishing filter taps of two consecutive low pass left boundary filters is 2.
7. The previous two statements hold for both left and right low and high pass boundary filters.

10.9 In (10.21) it is claimed that the result of transforming a N periodic signal $\mathbf{x}^p$ with $\mathbf{T}_a$ yields another N periodic signal $\mathbf{y}^p$.

1. Show that this is true, and that it is possible to reconstruct $\mathbf{x}$ using only $\mathbf{T}_s$ and $\mathbf{y}$.
2. Show that $\mathbf{y} = \mathbf{T}_a^p \mathbf{x}$ is the same signal as

$$\left[y[0]\ y[1]\ \cdots\ y[N-1] \right]^\top,$$

the period we selected in $\mathbf{y}^p = \mathbf{T}_a \mathbf{x}^p$.
3. Show that $\mathbf{T}_a^p$ in (10.22) is orthogonal.
   Hint: This is equivalent to showing that all the rows of $\mathbf{T}_a^p$ are mutually orthogonal and have norm 1.

10.10 Explain why it is correct to substitute $\tilde{g}[4] = h[4]$ (which eventually gave the 1 in (10.28)), despite the fact that $k = 2$ and $c = -1$ in (7.36).

11. Implementation

In order to implement the methods presented in the previous chapters some software package is obviously needed. We have chosen to use MATLAB, since it is well suited to perform signal processing. It is available for many platforms, and the programming language is independent of the platform. Although many things are explained in detail, we still assume some familiarity with MATLAB. If you have never used MATLAB, you should start by becoming familiar with the facilities, using the introduction shipped with MATLAB. We will also assume that you have worked through all the examples in Chap. 13, before you start on this chapter.

We assume that you have access to MATLAB. Some of the examples also provide C code for the implementation. To use these you should be familiar with this programming language and have access to a compiler. The C code examples are optional and can safely be ignored, unless you need to work on large data sets.

All the functions presented in this chapter are available electronically, see Chap. 14.

11.1 Introduction to Software

Let us start with basic things about the use of MATLAB. There are several ways to use MATLAB. The commands can be entered at the prompt, or a number of commands can be collected in a function. The advantage of using a function is that variables are local to that function, except those returned by the function call. A function is an ASCII file, with name extension .m, a so-called M-file. The first word in the file must be `function`. Function 1.1 gives an example.

Functions are very convenient in an implementation of a wavelet transform, since one can construct functions implementing a single step, and then use them to define the transform. This is illustrated in this chapter.

Another possibility is to write an M-file without function as the first word, containing a number of MATLAB commands. This is called a script. When the script is executed by typing its name at the prompt, the commands in the file are executed sequentially, as if they were typed at the MATLAB prompt. This also means that variables are not local. Scripts are convenient when testing programs consisting of several lines of MATLAB code. Adjustments can be made to the file, and all commands are easily re-executed, by typing the file name at the prompt again. Once everything works as it should, and you intend to reuse the code, you should turn it into a function.

MATLAB also offers the possibility to construct a graphical user interface to scripts and functions. This topic is not discussed here.

Let us look at Function 1.1 again. You should save it as `dwt.m` (note that this is just a simple example – the function does not implement a complete DWT). Concerning the name, then `dwt` might already be in use. This depends on your personal set-up of MATLAB. It is simple to check, if the name is already used. Give the command `dwt` at the prompt. If you receive an unknown function error message, then the name is not in use (obviously you should do this before naming the file).

In the first few examples the initial signal is contained in the vector `Signal`. Later we abbreviate the name to S.

Function 1.1 Example of a Function in MATLAB

```
function R = dwt(Signal)

N = length(Signal);      % Finds the length of the signal.

s = zeros(1,N/2);        % Predefines a vector of zeroes,
d = s;                   % and a copy of it.

% Here the signal is processed
% as in the following examples. See below.
% The result is placed in s and d.

R = [s d];               % Concatenates s and d.
```

It is important to remember that indexing in MATLAB starts with 1, and not 0. This is in contrast to the theory, where the starting index usually is 0. We will point out the changes necessary in the following examples.

11.2 Implementing the Haar Transform Through Lifting

We start by implementing the very first example from Chap. 2 as a MATLAB function. This example is given on p. 7, and for the reader's convenience we repeat the decomposition here in Table 11.1.

Although the function giving this decomposition is quite short, it is better to separate it in two functions. The first function calculates one step in the decomposition, and the second function then builds the wavelet decomposition using the one step function.

Table 11.1. The first decomposition

56	40	8	24	48	48	40	16
48	16	48	28	8	−8	0	12
32	38	16	10	8	−8	0	12
35	−3	16	10	8	−8	0	12

Let us assume that the signal is in `Signal`, and we want the means in the vector **s** and the differences in the vector **d**. The Haar transform equations are (see (3.1) and (3.2))

$$s = \frac{a+b}{2},$$
$$d = a - s.$$

Since MATLAB starts indexing at 1 we get

```
s(1) = 1/2*(Signal(1) + Signal(2));
d(1) = Signal(1) - s(1);
```

The next pair yields

```
s(2) = 1/2*(Signal(3) + Signal(4));
d(2) = Signal(3) - s(2);
```

and after two further computations **s** is the vector given by the first four numbers in the second row in the decomposition, and **d** is the last four numbers in the second row. This is generalized to a signal of length N, where we assume N is even.

```
for n=1:N/2
  s(n) = 1/2*(Signal(2*n-1) + Signal(2*n));
  d(n) = Signal(2*n-1) - s(n);
end
```

The function, shown in Function 2.1, could be named `dwthaar` (if this name is not already in use). It takes a vector (of even length) as input argument, and returns another vector of the same length, with the means in the first half and the differences in the second half. This is exactly what we need to construct the decomposition. Save this function in a file named `dwthaar.m`, and at the MATLAB prompt type

```
[s,d] = dwthaar([56 40 8 24 48 48 40 16])
```

This should give the elements of the second row in the decomposition.

Function 2.1 The Haar Transform

```
function [s,d] = dwthaar(Signal)

% Determine the length of the signal.
N = length(Signal);

% Allocate space in memory.
s = zeros(1, N/2);
d = s;

% The actual transform.
for n=1:N/2
  s(n) = 1/2*(Signal(2*n-1) + Signal(2*n));
  d(n) = Signal(2*n-1) - s(n);
end
```

Note that it is often a good idea to allocate the memory needed for the output vectors. This can be done by typing `s = zeros(1,Len)`, where Len is the desired length. Then MATLAB allocates all the memory at once instead of as it is needed. With short signals the time saved is minimal, but with a million samples a lot of time can be saved this way.

We now have a function, which can turn the first row in Table 11.1 into the second row, and the first four entries in the second row into the first four entries in the third row, and so on. The next step is to make a function, which uses `dwthaar` an appropriate number of times, to produce the entire decomposition. As input to this function we have again the signal, while the output is a matrix containing the decomposition, equivalent to Table 11.1. First the matrix T is allocated, then the signal is inserted as the first row. The `for` loop uses `dwthaar` to calculate the three remaining rows.

```
T = zeros(4,8);
T(1,:) = Signal;
for j=1:3
  Length = 2^(4-j);
  T(j+1, 1:Length) = dwthaar( T(j, 1:Length) );
  T(j+1, Length+1:8) = T(j, Length+1:8);
end
```

For each level the length of the elements, and hence the signal to be transformed, is determined. Since this length is halved for each increment of j (first 8, then 4, then 2), it is given as $2^(4-j) = 2^{4-j}$. Then the first part of the row is transformed, and the remaining part is copied to the next row. This piece of code can easily be extended to handle any signal of length 2^N for $N \in \mathbf{N}$. This is done with Function 2.2.

Function 2.2 Wavelet Decomposition Using the Haar Transform

```
function T = w_decomp(Signal)

N = size(Signal,2);
J = log2(N);

if rem(J,1)
  error('Signal must be of length 2^N.');
end

T = zeros(J, N);
T(1,:) = Signal;

for j=1:J
  Length = 2^(J+1-j);
  T(j+1, 1:Length) = dwthaar( T(j, 1:Length) );
  T(j+1, Length+1:N) = T(j, Length+1:N);
end
```

The variable J is the number of rows needed in the matrix. Since rem is the remainder after integer division, rem(J,1) is the fractional part of the variable. If it is not equal to zero, the signal is not of length 2^N, and an error message is displayed (and the program halts automatically).

Note that the representation in a table like Table 11.1 is highly inefficient and redundant due to the repetition of the computed differences. But it is convenient to start with it, in order to compare with the tables in Chap. 2.

11.3 Implementing the DWT Through Lifting

The implementation of the Haar transform was not difficult. This is the only transform, however, which is that easy to implement. This becomes clear in the following where we turn the attention to the Daubechies 4 transform. The problems we will encounter here apply to all wavelet transforms, when implemented as lifting steps. We will therefore examine in detail how to implement Daubechies 4, and in a later section we show briefly how to implement the transform CDF(4,6), which is rather complicated.

There are basically two different ways to implement lifting steps:

1. Each lifting step is applied to all signal samples (only possible when the entire signal is known).
2. All lifting steps are applied to each signal sample (always possible).

To see what this means, let us review the Daubechies 4 equations:

$$s^{(1)}[n] = S[2n] + \sqrt{3}S[2n+1] \,, \tag{11.1}$$

$$d^{(1)}[n] = S[2n+1] - \tfrac{1}{4}\sqrt{3}s^{(1)}[n] - \tfrac{1}{4}(\sqrt{3}-2)s^{(1)}[n-1] \,, \tag{11.2}$$

$$s^{(2)}[n] = s^{(1)}[n] - d^{(1)}[n+1] \,, \tag{11.3}$$

$$s[n] = \frac{\sqrt{3}-1}{\sqrt{2}}s^{(2)}[n] \,, \tag{11.4}$$

$$d[n] = \frac{\sqrt{3}+1}{\sqrt{2}}d^{(1)}[n] \,. \tag{11.5}$$

With the first method all the signal samples are 'sent through' equation (11.1). Then all the odd signal samples and all the $s^{(1)}$ samples are sent through equation (11.2), and so on. We will see this in more detail later.

With the second method the first equation is applied to first two signal samples, and the resulting $s^{(1)}$ is put into the next equation along with one odd signal sample and another $s^{(1)}$. Following this pattern we get one s and one d from (11.4) and (11.5). This is repeated with the third and fourth signal samples giving two more s and d values.

The first method is much easier to implement than the second one, especially in MATLAB. However, in a real application the second method might be the only option, if for example the transform has to take place while the signal is 'coming in.' For this reason we refer to the second method as the real time method, although it does not necessarily take place in real time. We start with the first method.

11.3.1 Implementing Daubechies 4

First we want to apply equation (11.1) to the entire signal, i.e. we want to calculate $s^{(1)}[n]$ for all values of n. For a signal of length N the calculation can be implemented in a for loop

```
for n=1:N/2
  s1(n) = S(2*n-1) + sqrt(3)*S(2*n);
end
```

Remember that MATLAB starts indexing at 1. Such a for loop will work, and it is easy to implement in other programs such as Maple, S-plus, or in the C language. But MATLAB offers a more compact and significantly faster solution. We can interpret (11.1) as a vector equation, where n goes from 1 to $N/2$. The equation is

$$
\begin{bmatrix} s^{(1)}[1] \\ s^{(1)}[2] \\ \vdots \\ s^{(1)}[N/2] \end{bmatrix} = \begin{bmatrix} S[1] \\ S[3] \\ \vdots \\ S[N-1] \end{bmatrix} + \sqrt{3} \begin{bmatrix} S[2] \\ S[4] \\ \vdots \\ S[N] \end{bmatrix} .
$$

Note that the indices correspond to MATLAB indexing. The vector equation becomes

```
s1 = S(1:2:N-1) + sqrt(3)*S(2:2:N);
```

in MATLAB code. This type of vector calculation is exactly what MATLAB was designed to handle, and this single line executes much faster than the `for` loop. The next equation is (11.2), and it contains a problem, since the value of $s^{(1)}[n-1]$ is not defined for $n = 0$. There are several different solutions to this problem (see Chap. 10). We choose periodization, since it gives a unitary transform and is easily implemented. In the periodized signal all undefined samples are taken from the other end of the signal, i.e. its periodic continuation. This means that we define $s^{(1)}[-1] \equiv s^{(1)}[N/2]$. Let us review equation (11.2) in vector form when we periodize.

$$\begin{bmatrix} d^{(1)}[1] \\ d^{(1)}[2] \\ \vdots \\ d^{(1)}[N/2] \end{bmatrix} = \begin{bmatrix} S[2] \\ S[4] \\ \vdots \\ S[N] \end{bmatrix} - \frac{\sqrt{3}}{4} \begin{bmatrix} s^{(1)}[1] \\ s^{(1)}[2] \\ \vdots \\ s^{(1)}[N/2] \end{bmatrix} - \frac{\sqrt{3}-2}{4} \begin{bmatrix} s^{(1)}[N/2] \\ s^{(1)}[1] \\ \vdots \\ s^{(1)}[N/2-1] \end{bmatrix}.$$

In MATLAB code this becomes

```
d1 = S(2:2:N) - sqrt(3)/4*s1 - (sqrt(3)-2)/4*[s1(N/2) s1(1:N/2-1)];
```

Again this vector implementation executes much faster than a `for` loop. Note how elegantly the periodization is performed. This would be more cumbersome with a `for` loop. The change of a vector from s1 to [s1(N/2) s1(1:N/2-1)] will in the following be referred to as a *cyclic permutation* of the vector s1.

It is now easy to complete the transform, and the entire transform is

Function 3.1 Matlab Optimized Daubechies 4 Transform

```
s1 = S(1:2:N-1) + sqrt(3)*S(2:2:N);
d1 = S(2:2:N) - sqrt(3)/4*s1 - (sqrt(3)-2)/4*[s1(N/2) s1(1:N/2-1)];
s2 = s1 - [d1(2:N/2) d1(1)];
s = (sqrt(3)-1)/sqrt(2) * s2;
d = (sqrt(3)+1)/sqrt(2) * d1;
```

This method for implementing lifting steps is actually quite easy, and since the entire signal is usually known in MATLAB, it is definitely the preferred one.

If implementing in another environment, for example in C, the vector operations might not be available, and the `for` loop suggested in the beginning of this section becomes necessary. The following function shows how to implement Function 3.1 in C.

Function 3.2 Daubechies 4 Transform in C

```
for (n = 0; n < N/2; n++) s[n] = S[2*n] + sqrt(3) * S[2*n+1];

d[0] = S[1] - sqrt(3)/4 * s[0] - (sqrt(3)-2)/4 * s[N/2-1];
for (n = 1; n < N/2; n++)
  d[n] = S[2*n+1] - sqrt(3)/4 * s[n] - (sqrt(3)-2)/4 * s[n-1];

for (n = 0; n < N/2-1; n++) s[n] = s[n] - d[n+1];
s[N/2-1] = s[N/2-1] - d[0];

for (n = 0; n < N/2; n++) s[n] = (sqrt(3)-1) / sqrt(2) * s[n];

for (n = 0; n < N/2; n++) d[n] = (sqrt(3)+1) / sqrt(2) * d[n];
```

Note that the periodization leads to two extra lines of code, one prior to the first loop, and one posterior to the second loop. The indexing now starts at 0. Consequently, the C implementation is closer to the Daubechies 4 equations (11.1) through (11.5) than to the MATLAB code.

There are a few things that can be improved in the above MATLAB and C implementations. We will demonstrate this in the following example.

11.3.2 Implementing CDF(4,6)

We now want to implement the CDF(4,6) transform, which is given by the following equations.

$$s^{(1)}[n] = S[2n] - \frac{1}{4}(S[2n-1] + S[2n+1]) , \tag{11.6}$$

$$d^{(1)}[n] = S[2n+1] - (s^{(1)}[n] + s^{(1)}[n+1]) , \tag{11.7}$$

$$s^{(2)}[n] = s^{(1)}[n] - \frac{1}{4096}\left(-35d^{(1)}[n-3] + 265d^{(1)}[n-2] - 998d^{(1)}[n-1]\right.$$
$$\left. - 998d^{(1)}[n] + 265d^{(1)}[n+1] - 35d^{(1)}[n+2]\right) , \tag{11.8}$$

$$s[n] = \frac{4}{\sqrt{2}}s^{(2)}[n] , \tag{11.9}$$

$$d[n] = \frac{\sqrt{2}}{4}d^{(1)}[n] . \tag{11.10}$$

This time the vector equations are omitted, and we give the MATLAB code directly. But first there are two things to notice.

Firstly, if we examine the MATLAB code from the Daubechies 4 implementation in Function 3.1, we see that there is really no need to use the three different variables s1, s2, and s. The first two variables might as well be changed to s, since there is no need to save s1 and s2.

Secondly, we will need a total of 7 different cyclically permuted vectors (with Daubechies 4 we needed 2). Thus it is preferable to implement cyclic permutation of a vector as a function. An example of such a function is

Function 3.3 Cyclic Permutation of a Vector

```
%
function P = cpv(S, k)

if k > 0
  P = [S(k+1:end) S(1:k)];
elseif k < 0
  P = [S(end+k+1:end) S(1:end+k)];
end
```

With this function we could write the second and third lines of the Daubechies 4 implementation in Function 3.1 as

```
d1 = S(2:2:N) - sqrt(3)/4*s1 - (sqrt(3)-2)/4*cpv(s1,-1);
s2 = s1 - cpv(d1,1);
```

With these two things in mind, we can now write a compact implementation of CDF(4,6).

Function 3.4 Matlab Optimized CDF(4,6) Transform

```
s = S(1:2:N-1) - 1/4*( cpv(S(2:2:N),-1) + S(2:2:N) );
d = S(2:2:N) - s - cpv(s,1);
s = s - 1/4096*( -35*cpv(d,-3) +265*cpv(d,-2) -998*cpv(d,-1) ...
                -998*d +265*cpv(d,1) -35*cpv(d,2) );
s = 4/sqrt(2) * s;
d = sqrt(2)/4 * d;
```

The three dots in the third line just tell MATLAB that the command continues on the next line. Typing the entire command on one line would work just as well (but this page is not wide enough for that!).

The C implementation of CDF(4,6) is shown in the next function. Notice that no less than five entries in s must be calculated outside one of the for loops (three before and two after).

Function 3.5 CDF(4,6) Transform in C

```
N = N/2;

s[0] = S[0] - (S[2*N-1] + S[1])/4;
for (n = 1; n < N; n++) s[n] = S[2*n] - (S[2*n-1] + S[2*n+1])/4;

for (n = 0; n < N-1; n++) d[n] = S[2*n+1] - (s[n] + s[n+1]);
d[N-1] = S[2*N-1] - (s[N-1] + s[0]);
```

```
s[0]  += (35*d[N-3]-265*d[N-2]+998*d[N-1]+998*d[0]-265*d[1]
          +35*d[2])/4096;
s[1]  += (35*d[N-2]-265*d[N-1]+998*d[0]+998*d[1]-265*d[2]
          +35*d[3])/4096;
s[2]  += (35*d[N-1]-265*d[0]+998*d[1]+998*d[2]-265*d[3]
          +35*d[4])/4096;

for (n = 3; n < N-2; n++)
  s[n]  += (35*d[n-3]-265*d[n-2]+998*d[n-1]+998*d[n]-265*d[n+1]
          +35*d[n+2])/4096;
s[N-2] += (35*d[N-5]-265*d[N-4]+998*d[N-3]+998*d[N-2]-265*d[N-1]
          +35*d[0])/4096;
s[N-1] += (35*d[N-4]-265*d[N-3]+998*d[N-2]+998*d[N-1]-265*d[0]
          +35*d[1])/4096;

K = 4/sqrt(2);
for (n = 0; n < N; n++) {
  s[n] *= K;
  d[n] /= K;
}
```

Due to the limited width of the page some of the lines have been split. Obviously, this is not necessary in an implementation.

11.3.3 The Inverse Daubechies 4 Transform

Daubechies 4Inverting the wavelet transform, i.e. implementing the inverse transform, is just as easy as the direct transform. We show only the inverse of Daubechies 4. Inverting CDF(4,6) is left as an exercise (and an easy one, too, see Exer. 11.3). As always when inverting a lifting transform the equations come in reverse order, and the variable to the left of the equal sign appears on the right side, and vice versa.

Function 3.6 Inverse Daubechies 4 Transform

```
d1 = d / ((sqrt(3)+1)/sqrt(2));
s2 = s / ((sqrt(3)-1)/sqrt(2));
s1 = s2 + cpv(d1,1);
S(2:2:N) = d1 + sqrt(3)/4*s1 + (sqrt(3)-2)/4*cpv(s1,-1);
S(1:2:N-1) = s1 - sqrt(3)*S(2:2:N);
```

11.4 The Real Time Method

When implementing according to the real time method we encounter a number of problems that do not exist for the first method. This is due to the fact that all transforms have references back and/or forth in time, for example the second and third equations in Daubechies 4 read

$$d^{(1)}[n] = S[2n + 1] - \frac{1}{4}\sqrt{3}s^{(1)}[n] - \frac{1}{4}(\sqrt{3} - 2)s^{(1)}[n - 1] , \qquad (11.11)$$

$$s^{(2)}[n] = s^{(1)}[n] - d^{(1)}[n + 1] , \qquad (11.12)$$

where $s^{(1)}[n - 1]$ refers to a previously computed value, and $d^{(1)}[n + 1]$ refers to a not yet computed value. Both types of references pose a problem, as will be clear in the next section. The advantage of the real time implementation is that it can produce two output coefficients (an s and a d value) each time two input samples are ready. Hence this method is suitable for a real time transform, which is a transformation of the signal as it becomes available.

11.4.1 Implementing Daubechies 4

We start by writing the equations in MATLAB code, as shown in Function 4.1. Note that, in contrast to previous functions, the scaling factor has its own variable K. The assignment of K is shown in this function, but omitted in subsequent functions to reduce the number of code lines.

Function 4.1 The Raw Daubechies 4 Equations

```
K = (sqrt(3)-1)/sqrt(2);
for n=1:N/2
  s1(n) = S(2*n-1) + sqrt(3)*S(2*n);
  d1(n) = S(2*n) - sqrt(3)/4*s1(n) - (sqrt(3)-2)/4*s1(n-1);
  s2(n) = s1(n) - d1(n+1);
  s(n) = s2(n) * K;
  d(n) = d1(n) / K;
end
```

Again we see the problem mentioned above. The most obvious one is d1(n+1) in the third line of the loop, since this value has not yet been computed. A less obvious problem is s1(n-1) in the second line of the loop. This is a previous calculated value, and as such is does not pose a problem. But in the very first run through the loop (for $n = 1$) we will need s1(0). Requesting this in MATLAB causes an error! The problem is easily solved by doing an initial computation before starting the loop.

Since the value d1(n+1) is needed in the third line, we could calculate that value in the previous line instead of d1(n). This means changing the second line to

```
d1(n+1) = S(2*(n+1)) - sqrt(3)/4*s1(n+1) - (sqrt(3)-2)/4*s1(n);
```

Now we no longer need s1(n-1). Instead we need s1(n+1), which can be calculated in the first line by changing it to

```
s1(n+1) = S(2*(n+1)-1) + sqrt(3)*S(2*(n+1));
```

The change in the two lines means that the loop never calculates s1(1) and d1(1), so we need to do this by hand. Note also that the loop must stop at $n = N/2 - 1$ instead of $n = N/2$, since otherwise the first two lines need undefined signal samples. This in turn means that the last three lines of the loop is not calculated for $n = N/2$, and this computation therefore also has to be done by hand.

Function 4.2 The Real Time Daubechies 4 Transform

```
s1(1) = S(1) + sqrt(3)*S(2);
s1(N/2) = S(N-1) + sqrt(3)*S(N);
d1(1) = S(2) - sqrt(3)/4*s1(1) - (sqrt(3)-2)/4*s1(N/2);

for n=1:N/2-1
  s1(n+1) = S(2*(n+1)-1) + sqrt(3)*S(2*(n+1));
  d1(n+1) = S(2*(n+1)) - sqrt(3)/4*s1(n+1) - (sqrt(3)-2)/4*s1(n);
  s2(n) = s1(n) - d1(n+1);
  s(n) = s2(n) * K;
  d(n) = d1(n) / K;
end

s2(N/2) = s1(N/2) - d1(1);
s(N/2) = s2(N/2) * K;
d(N/2) = d1(N/2) / K;
```

Notice how periodization is used when calculating d1(1) and s2(N/2). In the case of d1(1) it causes a problem, since it seems that we need S(N-1) and S(N), and they are not necessarily available. One solution would be to use another boundary correction method, but this would require somewhat more work to implement. Another solution is to shift the signal by two samples, as demonstrated in Function 4.3.

Function 4.3 The Real Time Daubechies 4 Shifted Transform

```
s1(1) = S(3) + sqrt(3)*S(4);
s1(N/2) = S(1) + sqrt(3)*S(2);
d1(1) = S(4) - sqrt(3)/4*s1(1) - (sqrt(3)-2)/4*s1(N/2);

for n=1:N/2-2
  s1(n+1) = S(2*(n+1)-1+2) + sqrt(3)*S(2*(n+1)+2);
  d1(n+1) = S(2*(n+1)+2) - sqrt(3)/4*s1(n+1) - (sqrt(3)-2)/4*s1(n);
  s2(n) = s1(n) - d1(n+1);
  s(n) = s2(n) * K;
  d(n) = d1(n) / K;
end

d1(N/2) = S(2) - sqrt(3)/4*s1(N/2) - (sqrt(3)-2)/4*s1(N/2-1);
s2(N/2-1) = s1(N/2-1) - d1(N/2);
```

```
s(N/2-1) = s2(N/2-1) * K;
d(N/2-1) = d1(N/2-1) / K;

s2(N/2) = s1(N/2) - d1(1);
s(N/2) = s2(N/2) * K;
d(N/2) = d1(N/2) / K;
```

Notice how one more loop has to be extracted to accommodate for this change.

Now, once the first four signal samples are available, Function 4.3 will produce the first two transform coefficients s(1) and d(1), and for each subsequent two signal samples available, another two transform coefficients can be calculated.

There is one major problem with the implementation in Function 4.3. It consumes a lot of memory, and more complex transforms will consume even more memory. The memory consumption is 5 times $N/2$, disregarding the memory needed for the signal itself. But it does not have to be that way. In reality, only two entries of each of the vectors s1, s2, and d1 are used in a loop. After that they are not used anymore. The simple solution is to change all s1 and s2 to s, and all d1 to d.

Function 4.4 The Memory Optimized Real Time Daubechies 4 Transform

```
s(1) = S(3) + sqrt(3)*S(4);
d(1) = S(4) - sqrt(3)/4*s(1) - (sqrt(3)-2)/4* (S(1) + sqrt(3)*S(2));

for n=1:N/2-2
  s(n+1) = S(2*(n+1)-1+2) + sqrt(3)*S(2*(n+1)+2);
  d(n+1) = S(2*(n+1)+2) - sqrt(3)/4*s(n+1) - (sqrt(3)-2)/4*s(n);
  s(n) = s(n) - d(n+1);
  s(n) = s(n) * K;
  d(n) = d(n) / K;
end

s(N/2) = S(1) + sqrt(3)*S(2);
d(N/2) = S(2) - sqrt(3)/4*s(N/2) - (sqrt(3)-2)/4*s(N/2-1);
s(N/2-1) = s(N/2-1) - d(N/2);
s(N/2-1) = s(N/2-1) * K;
d(N/2-1) = d(N/2-1) / K;

s(N/2) = s(N/2) - d(1);
s(N/2) = s(N/2) * K;
d(N/2) = d(N/2) / K;
```

In this case it does not cause any problems – the function still performs a Daubechies 4 transform. But it is not always possible just to drop the original indexing, as we shall see in the next section.

The inverse transform is, as always, easy to implement. It is shown in the following function.

Function 4.5 The Optimized Real Time Inverse Daubechies 4 Transform

```
d(N/2) = d(N/2) * K;
s(N/2) = s(N/2) / K;
s(N/2) = s(N/2) + d(1);

d(N/2-1) = d(N/2-1) * K;
s(N/2-1) = s(N/2-1) / K;
s(N/2-1) = s(N/2-1) + d(N/2);
S(2) = d(N/2) + sqrt(3)/4*s(N/2) + (sqrt(3)-2)/4*s(N/2-1);
S(1) = s(N/2) - sqrt(3)*S(2);

for n=N/2-2:-1:1
  d(n) = d(n) * K;
  s(n) = s(n) / K;
  s(n) = s(n) + d(n+1);
  S(2*(n+1)+2) = d(n+1) + sqrt(3)/4*s(n+1) + (sqrt(3)-2)/4*s(n);
  S(2*(n+1)-1+2) = s(n+1) - sqrt(3)*S(2*(n+1)+2);
end

S(4) = d(1) + sqrt(3)/4*s(1) + (sqrt(3)-2)/4* (S(1) + sqrt(3)*S(2));
S(3) = s(1) - sqrt(3)*S(4);
```

However, this function requires the signals s and d to be available 'backwards,' since the loop starts at $N/2$. Furthermore, it needs the value d(1) in the third line. It is of course possible to implement an inverse transform which transform from the beginning of the signal (instead of the end), and which requires only available samples. We will not show such a transform, but leave it as an exercise.

11.4.2 Implementing CDF(4,6)

As before, we start with the raw equations for CDF(4,6). Note that the scaling factor K = 4/sqrt(2) is omitted.

Function 4.6 The Raw CDF(4,6) Equations

```
for n = 1:N/2
  s1(n) = S(2*n-1) - 1/4*(S(2*n-2) + S(2*n));
  d1(n) = S(2*n) - s1(n) - s1(n+1);
  s2(n) = s1(n) - 1/4096*( -35*d1(n-3) +265*d1(n-2) -998*d1(n-1) ...
                          -998*d1(n) +265*d1(n+1) -35*d1(n+2) );
  s(n) = s2(n) * K;
  d(n) = d1(n) / K;
end
```

Obviously we need d1(n-3) through d1(n+2). We therefore change the second line in the loop to

```
d1(n+2) = S(2*(n+2)) - s1(n+2) - s1(n+3);
```

which this in turn leads us to change the first line to

```
s1(n+3) = S(2*(n+3)-1) - 1/4*(S(2*(n+3)-2) + S(2*(n+3)));
```

With these changes the loop counter must start with $n = 4$ (to avoid an error in the third line), and end with $N/2 - 3$ (to avoid an error in the first line). The loop now looks like

Function 4.7 The Modified CDF(4,6) Equations

```
for n = 4:N/2-3
  s1(n+3) = S(2*(n+3)-1) - 1/4*(S(2*(n+3)-2) + S(2*(n+3)));
  d1(n+2) = S(2*(n+2)) - s1(n+2) - s1(n+3);
  s2(n) = s1(n) - 1/4096*( -35*d1(n-3) +265*d1(n-2) -998*d1(n-1) ...
                          -998*d1(n) +265*d1(n+1) -35*d1(n+2) );
  s(n) = s2(n) * K;
  d(n) = d1(n) / K;
end
```

As before we are interested in optimizing memory usage. This time we have to be careful, though. The underlined `d1(n-1)` refers to the value `d1(n+2)` calculated in the second line three loops ago, and *not* the value `d(n)` calculated in the fifth line in the previous loop. This is obvious since d and d1 are two different variables. But if we just change the d1 to d in the second and third line the `d1(n-1)` (which then becomes `d(n-1)`) actually *will* refer to `d(n)` in the fifth line. The result is not a MATLAB error, but simply a transform different from the one we want to implement.

This faulty reference can be avoided by delaying the scaling in the fifth (and fourth) line. Since the oldest reference to d is `d(n-3)`, we need to delay the scaling with three samples. The last two lines of the loop then read

```
s(n-3) = s2(n-3) * K;
d(n-3) = d1(n-3) / K;
```

Notice that n-3 is precisely the lowest admissible value, since n starts counting at 4. This is not coincidental, since the value 4 was determined by the index of the oldest d, with index n-3.

We are now getting closer to a sound transform, but it still lacks the first three and last three runs through the loop. As before we need to do these by hand, and as before we need to decide what method we would like to use to handle the boundary problem. Above we saw how the periodization worked, and that this method requires samples from the other end of the signal (hence the name 'periodization'). There exists yet another method, which is not only quite elegant, but also has no requirement for samples from the other end of the signal.

The idea is very simple. Whenever we need to apply a step from the lifting building block (prediction or update step), which requires undefined samples, we choose a step from another lifting building block that does not require these undefined samples. If for example we want to apply

$$s[n] = s[n] - \frac{1}{4096}\Big(-35d[n-3] + 265d[n-2] - 998d[n-1]$$
$$- 998d[n] + 265d[n+1] - 35d[n+2]\Big)$$

for $n = 3$, we will need the undefined sample $d[0]$. Note that in the CDF equations here we have left out the enumeration, since this is the form they will have in the MATLAB implementation.

If we had chosen to periodize, we would use $d[N/2]$, and if we had chosen to zero pad, we would define $d[0] = 0$. But now we take another lifting building block, for example CDF(4,4), and use the second prediction step from this transform,

$$s[n] = s[n] - \frac{1}{128}\Big(5d[n-2] - 29d[n-1] - 29d[n] + 5d[n+1]\Big).$$

We will apply this boundary correction method to our current transform, and we will use CDF(1,1), CDF(4,2), and CDF(4,4), so we start by stating these (except for CDF(1,1), which is actually the Haar transform).

$$s[n] = S[2n] - \frac{1}{4}\big(S[2n-1] + S[2n+1]\big) \tag{11.13}$$

$$d[n] = S[2n+1] - (s[n] + s[n+1]) \tag{11.14}$$

$$\text{CDF}(4,2) \quad s[n] = s[n] - \frac{1}{16}\big(-3d[n-1] - 3d[n]\big) \tag{11.15}$$

$$\text{CDF}(4,4) \quad s[n] := s[n] - \frac{1}{128}\big(5d[n-2] - 29d[n-1]$$
$$- 29d[n] + 5d[n+1]\big) \tag{11.16}$$

$$\text{CDF}(4,6) \quad s[n] := s[n] - \frac{1}{4096}\big(-35d[n-3] + 265d[n-2] - 998d[n-1]$$
$$- 998d[n] + 265d[n+1] - 35d[n+2]\big) \tag{11.17}$$

$$d[n] = \frac{\sqrt{2}}{4}d[n] \tag{11.18}$$

$$s[n] = \frac{4}{\sqrt{2}}s[n] \tag{11.19}$$

The advantage of using transforms from the same family (in this case the CDF(4,x) family) is that the first two lifting steps and the scaling steps are the same.

First we examine the loop in Function 4.7 for $n = 3$. The first two lines cause not problems, but the third would require d(3-3), which is an undefined sample. Using CDF(4,4) instead, as suggested above, the first three lines read

```
s(3+3) = S(2*(3+3)-1) - 1/4*( S(2*(3+3)-2) + S(2*(3+3)) );
d(3+2) = S(2*(3+2)) - s(3+2) - s(3+3);
s(3)   = s(3) - 1/128*( 5*d(3-2) -29d(3-1) -29*d(3) +5*d(3+1) );
```

The smallest index of d is now 1 (instead of 0). For $n = 2$ and $n = 1$ we do the same thing, except use CDF(4,2) and CDF(1,1), respectively. We still need to calculate s(1) through s(3), and d(1), d(2). Of these only s(1) poses a problem, and once again we substitute another lifting step, namely the one from CDF(1,1). The transform know looks like

Function 4.8 The Modified CDF(4,6) Transform

```
s(1) = S(1) - S(2);                                                % CDF(1,1)

s(2) = S(2*2-1) - 1/4*(S(2*2-2) + S(2*2));                         % CDF(4,x)
d(1) = S(2*2-2) - s(1) - s(1+1);                                   % CDF(4,x)

s(3) = S(2*3-1) - 1/4*(S(2*3-2) + S(2*3));                         % CDF(4,x)
d(2) = S(2*3-2) - s(2) - s(2+1);                                   % CDF(4,x)

s(1+3) = S(2*(1+3)-1) - 1/4*( S(2*(1+3)-2) + S(2*(1+3)) );  % CDF(4,x)
d(1+2) = S(2*(1+2)) - s(1+2) - s(1+3);                             % CDF(4,x)
s(1) = s(1) + 1/2*d(1);                                            % CDF(1,1)

s(2+3) = S(2*(2+3)-1) - 1/4*( S(2*(2+3)-2) + S(2*(2+3)) );  % CDF(4,x)
d(2+2) = S(2*(2+2)) - s(2+2) - s(2+3);                             % CDF(4,x)
s(2) = s(2) - 1/16*( -3*d(1) -3*d(2) );                           % CDF(4,2)

s(3+3) = S(2*(3+3)-1) - 1/4*( S(2*(3+3)-2) + S(2*(3+3)) );  % CDF(4,x)
d(3+2) = S(2*(3+2)) - s(3+2) - s(3+3);                             % CDF(4,x)
s(3) = s(3) - 1/128*( 5*d(1) -29*d(2) -29*d(3) +5*d(4) );   % CDF(4,4)

for n = 4:N/2-3
  s(n+3) = S(2*(n+3)-1) - 1/4*( S(2*(n+3)-2) + S(2*(n+3)) );
  d(n+2) = S(2*(n+2)) - s(n+2) - s(n+3);
  s(n) = s(n) - 1/4096*( -35*d(n-3) +265*d(n-2) -998*d(n-1) ...
                        -998*d(n) +265*d(n+1) -35*d(n+2) );  % CDF(4,6)
  s(n-3) = s(n-3) * K;
  d(n-3) = d(n-3) / K;
end
```

When the same considerations are applied to the end of the signal (the last three run through of the loop), we get the final function.

Function 4.9 The Real Time CDF(4,6) Transform

```
s(1) = S(1) - S(2);                                                % CDF(1,1)
for n = 1:5
  s(n+1) = S(2*(n+1)-1) - 1/4*(S(2*(n+1)-2) + S(2*(n+1)));  % CDF(4,x)
  d(n) = S(2*n) - s(n) - s(n+1);                                   % CDF(4,x)
end

s(1) = s(1) + 1/2*d(1);                                            % CDF(1,1)
s(2) = s(2) - 1/16*( -3*d(1) -3*d(2) );                           % CDF(4,2)
s(3) = s(3) - 1/128*( 5*d(1) -29*d(2) -29*d(3) +5*d(4) );   % CDF(4,4)
```

```
for n = 4:N/2-3
  s(n+3) = S(2*(n+3)-1) - 1/4*( S(2*(n+3)-2) + S(2*(n+3)) );
  d(n+2) = S(2*(n+2)) - s(n+2) - s(n+3);
  s(n) = s(n) - 1/4096*( -35*d(n-3) +265*d(n-2) -998*d(n-1) ...
                         -998*d(n) +265*d(n+1) -35*d(n+2) );
  s(n-3) = s(n-3) * K;
  d(n-3) = d(n-3) / K;
end

d(N/2) = S(N) - s(N/2);                                      % CDF(1,1)

s(N/2-2) = s(N/2-2) - 1/4096*( -35*d(N/2-5) +265*d(N/2-4) ...
   -998*d(N/2-3) -998*d(N/2-2) +265*d(N/2-1) -35*d(N/2) ); % CDF(4,6)
s(N/2-1) = s(N/2-1) - 1/128*( 5*d(N/2-3) -29*d(N/2-2) ...
                              -29*d(N/2-1) +5*d(N/2) );      % CDF(4,4)
s(N/2) = s(N/2) - 1/16*( -3*d(N/2-1) -3*d(N/2) );           % CDF(4,2)

for k=5:-1:0
  s(N/2-k) = s(N/2-k) * K;
  d(N/2-k) = d(N/2-k) / K;
end
```

Some of the lines have been rearranged in comparison to Function 4.8, in order to reduce the number of code lines. The values s(1) through s(6) and d(1) through d(5) might as well be calculated in advance, which is easier, since then we can use a for loop.

Since the real time method transforms sample by sample, and hence is expressed in terms of a for loop (even in MATLAB code), it is easy to convert Function 4.9 to C. It is simply a matter of changing the syntax and remember to start indexing at 0.

Function 4.10 The Real Time CDF(4,6) Transform in C

```
s[0] = S[0] - S[1];
for (n = 0; n < 5; n++) {
  s[n+1] = S[2*(n+1)] - (S[2*(n+1)-1] + S[2*(n+1)+1])/4;
  d[n] = S[2*n+1] - s[n] - s[n+1];
}

s[0] += d[0]/2;
s[1] -= (-3*d[0]-3*d[1])/16;
s[2] -= (5*d[0]-29*d[1]-29*d[2]+5*d[3])/128;

for (n = 3; n < N/2-3; n++) {
  s[n+3] = S[2*(n+3)] - (S[2*(n+3)-1] + S[2*(n+3)+1])/4;
  d[n+2] = S[2*(n+2)+1] - s[n+2] - s[n+3];
  s[n] += (35*d[n-3]-265*d[n-2]+998*d[n-1]+998*d[n]-265*d[n+1]
          +35*d[n+2])/4096;
  s[n-3] = s[n-3] * K;
  d[n-3] = d[n-3] / K;
```

```
}

N = N/2;
d[N-1] = S[2*N-1] - s[N-1];

s[N-3] += (35*d[N-6]-265*d[N-5]+998*d[N-4]+998*d[N-3]-265*d[N-2]
           +35*d[N-1])/4096;
s[N-2] -= ( 5*d[N-4] -29*d[N-3] -29*d[N-2]  +5*d[N-1])/128;
s[N-1] -= (-3*d[N-2]   -3*d[N-1])/16;

for (n = 6; n > 0; n--) {
  s[N-n] *= K;
  d[N-n] /= K;
}
```

The inverse of the transform is once again easy to implement, in MATLAB as well as in C. Here it is shown in MATLAB code.

Function 4.11 The Real Time Inverse CDF(4,6) Transform

```
for k=0:5
  d(N/2-k) = d(N/2-k) * K;
  s(N/2-k) = s(N/2-k) / K;
end

s(N/2) = s(N/2) + 1/16*( -3*d(N/2-1) -3*d(N/2) );
s(N/2-1) = s(N/2-1) + 1/128*( 5*d(N/2-3) -29*d(N/2-2) ...
            -29*d(N/2-1) +5*d(N/2) );
s(N/2-2) = s(N/2-2) + 1/4096*( -35*d(N/2-5) +265*d(N/2-4) ...
            -998*d(N/2-3) -998*d(N/2-2) +265*d(N/2-1) -35*d(N/2) );
S(N) = d(N/2) + s(N/2);

for n = N/2-3:-1:4
  d(n-3) = d(n-3) * K;
  s(n-3) = s(n-3) / K;
  s(n) = s(n) + 1/4096*( -35*d(n-3) +265*d(n-2) -998*d(n-1) ...
          -998*d(n) +265*d(n+1) -35*d(n+2) );
  S(2*(n+2)) = d(n+2) + s(n+2) + s(n+3);
  S(2*(n+3)-1) = s(n+3) + 1/4*(S(2*(n+3)-2) + S(2*(n+3)));
end

s(3) = s(3) + 1/128*( 5*d(1) -29*d(2) -29*d(3) +5*d(4));
s(2) = s(2) + 1/16*( -3*d(1) -3*d(2));
s(1) = s(1) - 1/2*d(1);

for n = 5:-1:1
  S(2*n) = d(n) + s(n) + s(n+1);
  S(2*(n+1)-1) = s(n+1) + 1/4*(S(2*(n+1)-2) + S(2*(n+1)));
end
S(1) = s(1) + S(2);
```

We finish the implementation of the wavelet transform through lifting by showing an optimized version of the real time CDF(4,6) implementation. Here we take advantage of the fast vector operations available in MATLAB.

Function 4.12 The Optimized Real Time CDF(4,6) Transform.

```
N = length(S)/2;
cdf2 = 1/16 * [-3 -3];
cdf4 = 1/128 * [5 -29 -29 5];
cdf6 = 1/4096 * [-35 265 -998 -998 265 -35];

s(1) = S(1) - S(2);                           % CDF(1,1)
s(2:6) = S(3:2:11) - (S(2:2:10) + S(4:2:12))/4;  % CDF(4,x)
d(1:5) = S(2:2:10) - s(1:5) - s(2:6);         % CDF(4,x)

s(1) = s(1) + d(1)/2;                         % CDF(1,1)
s(2) = s(2) - cdf2 * d(1:2)';                 % CDF(4,2)
s(3) = s(3) - cdf4 * d(1:4)';                 % CDF(4,4)

for n = 4:N-3
  s(n+3) = S(2*n+5) - (S(2*n+4) + S(2*n+6))/4;
  d(n+2) = S(2*n+4) - s(n+2) - s(n+3);
  s(n) = s(n) - cdf6 * d(n-3:n+2)';
  s(n-3) = s(n-3) * K;
  d(n-3) = d(n-3) / K;
end

d(N) = S(2*N) - s(N);                         % CDF(1,1)

s(N-2) = s(N-2) - cdf6 * d(N-5:N)';           % CDF(4,6)
s(N-1) = s(N-1) - cdf4 * d(N-3:N)';           % CDF(4,4)
s(N) = s(N) - cdf2 * d(N-1:N)';               % CDF(4,2)

s(N-5:N) = s(N-5:N) * K;
d(N-5:N) = d(N-5:N) / K;
```

11.4.3 The Real Time DWT Step-by-Step

In the two previous sections we have shown how to implement the Daubechies 4 and the CDF(4,6) transforms. In both cases the function implementing the transform consists of a core, in the form of a `for` loop, which performs the main part of the transformation, and some extra code, which handles the boundaries of the signal. While there are many choices for a boundary handling method (two have been explored in the previous sections), the core of the transform always has the same structure.

Based on the two direct transforms, Daubechies 4, implemented in Function 4.4, and CDF(4,6), implemented in Function 4.9, we give an algorithm for a real time lifting implementation of any transform.

1. Write the raw equations in a `for` loop, which runs from 1 to $N/2$ (see Function 4.1).
2. Remove any suffixes by changing s1, s2, and so on to s, and likewise with d.
3. Start with the last equation (except for the two scale equations) and find the largest index, and then change the previous equation accordingly. For example with CDF(4,6)

```
d(n) = S(2*n) - s(n) - s(n+1);
s(n) = s(n) - 1/4096*( -35*d(n-3) +265*d(n-2) -998*d(n-1) ...
                       -998*d(n) +265*d(n+1) -35*d(n+2) );
```

is changed to

```
d(n+2) = S(2*(n+2)) - s(n+2) - s(n+3);
s(n) = s(n) - 1/4096*( -35*d(n-3) +265*d(n-2) -998*d(n-1) ...
                       -998*d(n) +265*d(n+1) -35*d(n+2) );
```

If the largest index is less than n, the previous equation should *not* be changed.
4. Do this for all the equations, ending with the first equation.
5. Find the smallest and largest index in all of the equations, and change the `for` loop accordingly. For example in CDF(4,6) the smallest index is d(n-3) and the largest s(n+3). The loop is then changed to (see Function 4.7)

```
for n=4:N/2-3
```

6. Change the two scaling equations such that they match the smallest index. In CDF(4,6) this is n-3, and the scaling equations are changed to

```
s(n-3) = s2(n-3) * K;
d(n-3) = d1(n-3) / K;
```

7. Finally, apply some boundary handling method to the remaining indices. For CDF(4,6) this would be $n = 1, 2, 3$ and $n = N/2 - 2, N/2 - 1, N/2$.

11.5 Filter Bank Implementation

The traditional implementation of the DWT is as a filter bank. The filters were presented in Chap. 7, but without stating the implementation formula. The main difference between the lifting implementation and the filter bank implementation is the trade-off between generality and speed. The lifting approach requires a new implementation for each transform (we implemented Daubechies 4 and CDF(4,6) in the previous section, and they were quite different), whereas the filter bank approach has a fixed formula, independent of the transform. A disadvantage is that filtering, as a rule of thumb, requires twice as many calculations as lifting. In some applications the generality is more important than speed and we therefore also present briefly implementation of filtering.

Implementation of the wavelet transform as a filter bank is discussed many places in the wavelet literature. We have briefly discussed this topic a number of times in the previous chapters, and we will not go into further detail here. Instead we will show one implementation of a filter bank. Readers interested in more detailed information are referred to Wickerhauser [30] and the available C code (see Chap. 14).

11.5.1 An Easy MATLAB Filter Bank DWT

The following MATLAB function is the main part of the *Uvi_Wave* function wt.m, which performs the wavelet transform. It takes the signal, a filter pair, and the number of scales as input, and returns the transformed signal. The function uses the MATLAB function conv (abbreviation of convolution which is filtering) to perform the low/high pass filtering, and then subsequently decimates the signal (down sample by two). This is a very easy solution (and actually corresponds to the usual presentation of filter banks, see for example Fig. 7.4), but it is also highly inefficient, since calculating a lot of samples, just to use half of them, is definitely not the way to do a good implementation. Note that there is no error checking in the function shown here (there is in the complete wt.m from *Uvi_Wave*, though), so for instance a too large value of k (more scales than the length of the signal permits) or passing S as a column vector will not generate a proper error. The structure of the output vector R is described in help wt. The variables dlp and dhp are used to control the alignment of the output signal. This is described in more detail in Sect. 9.4.3 on p. 116.

Function 5.1 Filter implementation of DWT (*Uvi_Wave*)

```
function R = fil_cv(S,h,g,k)

% Copyright (C) 1994, 1995, 1996, by Universidad de Vigo

llp = length(h);          % Length of the low pass filter
lhp = length(g);          % Length of the high pass filter.

L = max([lhp,llp]);       % Number of samples for the wraparound.

% Start the algorithm
R = [];                   % The output signal is reset.

for i = 1:k               % For every scale (iteration)...
  lx = length(S);
```

```
if rem(lx,2) ~= 0          % Check that the number of samples
   S = [S,0];              % will be even (because of decimation).
   lx = lx + 1;
end

Sp = S;                    % Build wraparound. The input signal
pl = length(Sp);           % can be smaller than L, so it may
while L > pl               % be necessary to repeat it several
   Sp = [Sp,S];            % times.
   pl = length(Sp);
end

S = [Sp(pl-L+1:pl),S,Sp(1:L)];   % Add the wraparound.

s = conv(S,h);             % Then do low pass filtering
d = conv(S,g);             % and high pass filtering.

s = s((1+L):2:(L+lx));     % Decimate the outputs
d = d((1+L):2:(L+lx));     % and leave out wraparound

R = [d,R];                 % Put the resulting wavelet step
                           % on its place in the wavelet vector,
S = s;                     % and set the next iteration.
end

R = [S,R];                 % Wavelet vector (1 row vector)
```

The word 'wraparound' is equivalent to what we in this book prefer to denote 'periodization.' This principle is described in Sect. 10.4.

11.5.2 A Fast C Filter Bank DWT

The filter bank implementation is well suited for the C language. It is more efficient to use pointers than indices in C, and the structure of the filter bank transform makes it easy to do just that. The following function demonstrates how pointers are used to perform the transform. In this case we have chosen to use boundary filters instead of periodization. The method of boundary filters is presented in Sect. 10.3. The construction of these boundary filters is shown later in this chapter, in Sect. 11.6 below.

To interpret this function a certain familiarity with the C language is required, since we make no attempt to explain how it works. The reason for showing this function is that an efficient C implementation typically transforms 1.000 to 10.000 times faster than the various *Uvi_Wave* transform implementations.

In this function N is the length of the signal, HLen the length of the ordinary filters (only orthogonal filters can be used in this function). HA, GA, and LBM are pointers to the ordinary and boundary filters, respectively. Finally, EMN is the number of boundary filters at each end. In contrast to the previously presented transform implementation in this chapter, this function includes Gray code permutation (see Sect. 9.3). This piece of code is from dwte.c, which is available electronically, see Chap. 14.

Function 5.2 Filter implementation with boundary correction – in C

```
double *SigPtr, *SigPtr1, *SigPtr2, *Hptr, *Gptr, *BM;
int GCP, EndSig, m, n;

if (fmod(HLen,4)) GCP = 0; else GCP = 1;

SigPtr1 = &RetSig[GCP*N/2];
SigPtr2 = &RetSig[(1-GCP)*N/2];

/* LEFT EDGE CORRECTION (REALLY A MUL OF MATRIX AND VECTOR). */
BM = LBM;
for (n = 0; n < EMN-1; n += 2)
{
  SigPtr = Signal;
  *SigPtr1 = *BM++ * *SigPtr++;
  for (m = 1; m < EMM-1; m++) *SigPtr1 += *BM++ * *SigPtr++;
  *SigPtr1++ += *BM++ * *SigPtr++;

  SigPtr = Signal;
  *SigPtr2 = *BM++ * *SigPtr++;
  for (m = 1; m < EMM-1; m++) *SigPtr2 += *BM++ * *SigPtr++;
  *SigPtr2++ += *BM++ * *SigPtr++;
}

if (!fmod(HLen,4))
{
  SigPtr = Signal;
  *SigPtr1 = *BM++ * *SigPtr++;
  for (m = 1; m < EMM-1; m++) *SigPtr1 += *BM++ * *SigPtr++;
  *SigPtr1++ += *BM++ * *SigPtr++;

  SigPtr = SigPtr1;
  SigPtr1 = SigPtr2;
  SigPtr2 = SigPtr;
}

/* THE ORDINARY WAVELET TRANSFORM (ON THE MIDDLE OF THE SIGNAL). */
for (n = 0; n < N/2-EMN; n++)
{
  SigPtr = &Signal[2*n];
  Hptr = HA;
  Gptr = GA;

  *SigPtr1 = *Hptr++ * *SigPtr;
  *SigPtr2 = *Gptr++ * *SigPtr++;
  for (m = 1; m < HLen-1; m++)
  {
    *SigPtr1 += *Hptr++ * *SigPtr;
```

```
        *SigPtr2 += *Gptr++ * *SigPtr++;
    }
    *SigPtr1++ += *Hptr++ * *SigPtr;
    *SigPtr2++ += *Gptr++ * *SigPtr++;
}

/* RIGHT EDGE CORRECTION (REALLY A MUL OF MATRIX AND VECTOR). */
EndSig = N-EMM;
BM = RBM;
for (n = 0; n < EMN-1; n += 2)
{
    SigPtr = &Signal[EndSig];
    *SigPtr1 = *BM++ * *SigPtr++;
    for (m = 1; m < EMM-1; m++) *SigPtr1 += *BM++ * *SigPtr++;
    *SigPtr1++ += *BM++ * *SigPtr++;

    SigPtr = &Signal[EndSig];
    *SigPtr2 = *BM++ * *SigPtr++;
    for (m = 1; m < EMM-1; m++) *SigPtr2 += *BM++ * *SigPtr++;
    *SigPtr2++ += *BM++ * *SigPtr++;
}

if (!fmod(HLen,4))
{
    SigPtr = &Signal[EndSig];
    *SigPtr1 = *BM++ * *SigPtr++;
    for (m = 1; m < EMM; m++) *SigPtr1 += *BM++ * *SigPtr++;
}
```

The disadvantage of using pointers instead of indices is that the code becomes difficult to read. This can be counteracted by inserting comments in the code, but it would make the function twice as long. We have chosen to omit comments in this function, and simply let it illustrate what a complete and optimized implementation of a filter bank DWT looks like in C. The original function dwte has many comments inserted.

11.6 Construction of Boundary Filters

There are may types of boundary filters, and we have in Chap. 10 presented two types, namely the ones we called Gram-Schmidt boundary filters, and those that preserve vanishing moments. Both apply, as presented, only to orthogonal filters. The first method is quite easy to implement, while the second method is more complicated, and it requires a substantial amount of computation.

Because of the complicated procedure needed in the vanishing moments case, we will omit this part (note that a MATLAB file is electronically available), and limit the implementation to Gram-Schmidt boundary filters.

11.6.1 Gram-Schmidt Boundary Filters

The theory behind this method is presented in Sect. 10.3, so in this section we focus on the implementation issues only. We will implement the method according to the way the $\mathbf{M}$ and $\mathbf{M}'$ matrices in (10.19) and (10.20) were constructed. The only difference is that we will omit the ordinary filters in the middle of the matrices, since they serve no purpose in this context.

We start the construction with a matrix containing all possible even truncations of the given filters.

$$\begin{bmatrix} h[1] & h[0] & \cdots & 0 & 0 & 0 & 0 \\ g[1] & g[0] & \cdots & 0 & 0 & 0 & 0 \\ \vdots & \vdots & \ddots & \vdots & \vdots & \vdots & \vdots \\ h[L-5] & h[L-6] & \cdots & h[1] & h[0] & 0 & 0 \\ g[L-5] & g[L-6] & \cdots & g[1] & g[0] & 0 & 0 \\ h[L-3] & h[L-4] & \cdots & h[3] & h[2] & h[1] & h[0] \\ g[L-3] & g[L-4] & \cdots & g[3] & g[2] & g[1] & g[0] \end{bmatrix}. \qquad (11.20)$$

The number of left boundary filters is $L/2 - 1$, so we reduce the matrix to the bottom $L/2 - 1$ rows (which is exactly half of the rows). Note that the first row in the reduce matrix is (part of) the low pass filter, if $L = 4K + 2$, but (part of) the high pass filter, if $L = 4K$. Consequently, the last row is always a high pass filter. The next step is to Gram-Schmidt orthogonalize the rows, starting with the first row. Finally, the rows are normalized. The same procedure is used to construct the right boundary filters.

Function 6.1 Construction of Left and Right Boundary Filters

```
function [LBM, RBM] = boundary(H);

H = H(:)';                                    % Ensures a row vector
L = length(H);
G = fliplr(H).*((-1).^[0:L-1]);               % Construct high pass

% Construct matrices from H and G.
for k = 2:2:L-2
  LBM(k-1,1:k)       = H(L-k+1:L);            % Construct left boundary matrix
  LBM(k  ,1:k)       = G(L-k+1:L);
  RBM(k-1,L-k-1:L-2) = G(1:k);                % Construct right boundary matrix
  RBM(k  ,L-k-1:L-2) = H(1:k);                % which is upside down
end

LBM = LBM(L/2:L-2,:);                         % Truncate to last half of rows
RBM = RBM(L/2:L-2,:);

% Do Gram-Schmidt on rows of LBM.
for k = 1:L/2-1
  v = LBM(k,:) - (LBM(1:k-1,:) * LBM(k,:)')' * LBM(1:k-1,:);
  LBM(k,:) = v/norm(v);
```

```
end

% Do Gram-Schmidt on rows of RBM.
for k = 1:L/2-1
   v = RBM(k,:) - (RBM(1:k-1,:) * RBM(k,:)')')' * RBM(1:k-1,:);
   RBM(k,:) = v/norm(v);
end

RBM = flipud(RBM);                    % Flip right matrix upside down
```

The first for loop constructs the left matrix, as shown in (11.20), and the right matrix, followed by a truncation to the last half of the rows. Note that RBM is upside down. Then the Gram-Schmidt procedure is applied. Here we take advantages of MATLAB's ability to handle matrices and vectors to do the sum required in the procedure (see (10.14)). Of course, the sum can also be implemented as a for loop, which would be the only feasible way in most programming environments. The normalization is done after each orthogonalization.

This function only calculates the analysis boundary filters, but since they are constructed to give an orthogonal transform, the synthesis boundary filters are found simply by transposing the analysis boundary filters.

It is relatively simple to use the Gram-Schmidt boundary filters. The matrices constructed with Function 6.1 is multiplied with the ends of the signal, and the interior filters are applied as usual. To determine exactly where the boundary and interior filters are applied, we can make use of the transform in matrix form, as discussed in Chap. 10.

It is a bit more tricky to do the inverse transformation of the signal. First we note that the matrix form in for example (10.12) results in a mixing of the low and high pass transform coefficients, as described in Sect. 10.2. We therefore have two options. Either we separate the low and high pass parts prior to applying the inverse transform (the inverse transform then has the structure known from the first several chapters), or we apply a slightly altered transform, which fits the mixing of low and high pass coefficients. In the former case we use the ordinary synthesis filters, and since the synthesis boundary filters are given as the transpose of the analysis boundary matrices LBM and RBM, they have to be separated, too. In the latter case the ordinary synthesis filters do not apply immediately (the boundary filters do, though). We will focus on the latter case, leaving the former as an exercise.

To see what the inverse transform looks like for a transformed signal with mixed low and high pass coefficients, we first turn to (10.9). We see that the synthesis filters are columns of the transform matrix, but when that matrix is applied to a signal, the inner products, the filtering, happens row-wise. Examining the two full rows, however, we see that the low and high pass filters coefficients are mixed, which corresponds nicely to the fact that the signal is also a mix of the low and high pass parts. Therefore, if we use the two full rows in (10.9) as filters, we get a transform which incorporates both up sampling and addition in the filters. At least the addition is usually a separate action in a filter bank version of the inverse transform (see for instance Fig. 7.2 and Fig. 7.4).

The matrix of inverse transform is given as the transpose of the direct transform matrix. The synthesis matrix is shown in (10.9) before truncation. The analysis filters occur vertically in the matrix, but we show them horizontally below. For instance, if the analysis filters are given by

$$\mathbf{h} = \begin{bmatrix} h[1] & h[2] & h[3] & h[4] & h[5] & h[6] \end{bmatrix},$$
$$\mathbf{g} = \begin{bmatrix} g[1] & g[2] & g[3] & g[4] & g[5] & g[6] \end{bmatrix},$$

then the new, horizontal filters are

$$\mathbf{h}_r = \begin{bmatrix} h[5] & g[5] & h[3] & g[3] & h[1] & g[1] \end{bmatrix},$$
$$\mathbf{g}_r = \begin{bmatrix} h[6] & g[6] & h[4] & g[4] & h[2] & g[2] \end{bmatrix}.$$

Note also how these new filters are used both as whole filters and as truncated filters.

This alternative implementation of the synthesis matrix is also illustrated in Fig. 11.1. The analysis matrix is given as a number of interior (whole) filters and the left and right boundary filters. Here the boundary filters are shown as two matrices. The inverse transform matrix is the transpose of the analysis transform matrix, and when we consider the transposed matrix as consisting of filters row-wise, the structure of the synthesis matrix is as shown in the left matrix in Fig. 11.1. The boundary filters are still given as two submatrices, which are the transpose of the original boundary filter matrices. This figure is also useful in understanding the implementation of the direct and inverse transform in the filter bank version. The Function 6.2 demonstrates the use of the boundary filters to transform and inversely transform a signal.

Function 6.2 Application of Gram-Schmidt Boundary Filters

```
function S2 = ApplyBoundary(S)

L = 10;                          % The length of the filter
[h,g] = daub(L);                 % Get Daubechies L filter
[LBM,RBM] = boundary(h);         % Construct GS boundary filters

% Construction of alternative synthesis filters
hr(1:2:L-1) = h(L-1:-2:1);
hr(2:2:L) = g(L-1:-2:1);
gr(1:2:L-1) = h(L:-2:2);
gr(2:2:L) = g(L:-2:2);

N = length(S);

% Initialize transform signals
T = zeros(1,N);
```

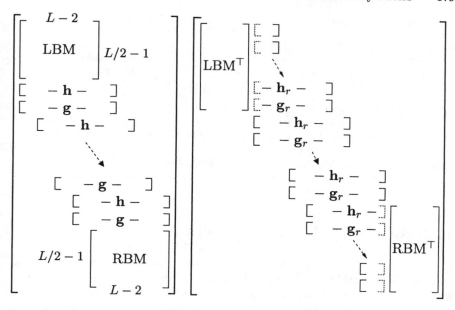

Fig. 11.1. The direct (left) transform matrix is constructed by the left and right boundary matrices and a number of ordinary, whole filters. The inverse (right) transform matrix is the transpose of the direct transform matrix, and the structure shown here is a result of interpreting the matrix as consisting of filters row-wise

```
T2 = T;
S2 = T2;

% Direct transform with GS boundary filters
T(1:L/2-1)   = LBM * S(1:L-2)';          % Apply left matrix
T(N-L/2+2:N) = RBM * S(N-L+3:N)';         % Apply right matrix

for k = 1:2:N-L+1
  T(k+L/2-1) = h * S(k:L+k-1)';           % Apply interior filters
  T(k+L/2)   = g * S(k:L+k-1)';
end

T = [T(1:2:N-1) T(2:2:N)];                % Separate low and high pass


% Inverse transform with GS boundary filters
T2(1:2:N-1) = T(1:N/2);                   % Mix low and high pass
T2(2:2:N)   = T(N/2+1:N);

for k = 1:2:L-3
  S2(k)   = hr(L-k:L) * T2(L/2:L/2+k)';    % Apply truncated
  S2(k+1) = gr(L-k:L) * T2(L/2:L/2+k)';    %   interior filters
end
```

```
S2(1:L-2) = S2(1:L-2) + (LBM' * T2(1:L/2-1)')'; % Apply left matrix

for k = 1+(L/2-1):2:N-L+1-(L/2-1)
   S2(k+L/2-1) = hr * T2(k:L+k-1)';                % Apply whole
   S2(k+L/2)   = gr * T2(k:L+k-1)';                %  interior filters
end

for k = N-L+3:2:N-1
   S2(k)   = hr(1:N-k+1) * T2(k-L/2+1:N-L/2+1)'; % Apply truncated
   S2(k+1) = gr(1:N-k+1) * T2(k-L/2+1:N-L/2+1)'; %  interior filters
end
S2(N-L+3:N) = S2(N-L+3:N) + (RBM' * T2(N-L/2+2:N)')'; % Right matrix
```

11.7 Wavelet Packet Decomposition

In most applications one DWT is not enough, and often it is necessary to do a complete wavelet packet decomposition. This means applying the DWT several times to various signals. The wavelet packets method was presented in Chap. 8. Here we show how to implement this method. We need to construct a function, which takes as input a signal, two filters, and the number of levels in the decomposition, and returns the decomposition in a matrix. It is just a matter of first applying the DWT to one signal, then to two signal, then to four signal, an so on.

Function 7.1 Wavelet Packet Decomposition

```
function D = wpd(S,h,g,J)

N = length(S);

if J > floor(log2(N))
  error('Too many levels.');
elseif rem(N,2^(J-1))
  error(sprintf('Signal length must be a multiple of 2^%i.',J-1));
end

D = zeros(J,N);
D(1,:) = S;

% For each level in the decomposition
% (starting with the second level).
for j = 1:J-1
  width = N/2^(j-1);        % Width of elements on j'th level.

  % For each pair of elements on the j'th level.
  for k = 1:2^(j-1)
    Interval = [1+(k-1)*width:k*width];
    D(j+1,Interval) = dwt(D(j,Interval),h,g);
```

```
     end
end
```

There are two loops in this function. One for the levels in the decomposition, and one for the elements on each level. Alternatively, the dwt function could be made to handle more than one element at the time. The electronically available function dwte takes a number of signals as input, and transforms each of them. Thus the two for loops in the previous function can be reduced to

```
for j=1:J-1
  D(j+1,:) = dwte(D(j,:),h,g,2^(j-1));
end
```

The fourth argument gives the number of signals within the signal D(j,:).

11.8 Wavelet Packet Bases, Basis Representation, and Best Basis

Once the full wavelet packet decomposition has been computed to a prescribed level (compatible with the length of the signal), a possible next step is to find the best basis. Implementing the best basis algorithm is not difficult, but there is one point which needs to be settled before we can proceed. We have to decide how to represent a basis in the computer. We will focus on MATLAB, but the principle applies to all types of software. This issue is addressed in the first section. The following two sections discuss implementation of cost computation and of best basis search.

11.8.1 Basis Representation

There are two different ways to represent a basis in a wavelet packet decomposition in MATLAB. In *Uvi_Wave* the basis is described by a binary tree, and the basis is represented by the depth of the terminal nodes, starting from the lowest frequency node, located on the left. See the left hand part of Fig. 11.2. The other representation is given by going through the entire binary tree, marking selected nodes with 1 and unselected nodes with 0, starting from the top, counting from left to right. This principle is also shown in Fig. 11.2 In both cases the representation is the vector containing the numbers described above.

The choice of representation is mostly a matter of taste, since they both have advantages and disadvantages. In this book we have chosen the second representation, and the following MATLAB functions are therefore based on this representation. A conversion function between the two representations is available electronically, see Chap. 14.

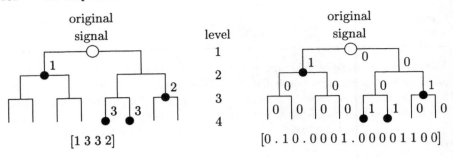

Fig. 11.2. Two different basis representation. Either we write the level of each element, starting from the left, or we go through the elements from top to bottom, marking the selected elements with 1 and unselected elements with 0

11.8.2 Calculating Cost Values for the Decomposition

In a decomposition with J levels there will be a total of

$$2^0 + 2^1 + \cdots + 2^{J-1} = 2^J - 1$$

elements, and we enumerate them as shown in Fig. 11.3. The best basis search

1							
2				3			
4		5		6		7	
8	9	10	11	12	13	14	15

Fig. 11.3. The enumeration of the elements in a full wavelet decomposition with four levels

uses two vectors, each of length $2^J - 1$. One contains the cost values for each elements, and the other contains the best basis, once we have completed the basis search.

In the following we assume that D is a matrix containing the wavelet packet decomposition of a signal S of length 2^N, down to level J. This decomposition could for example be the output from Function 7.1. An element in this decomposition is a vector consisting of some of the entries of a row in the matrix D. We will need to extract all the elements in the decomposition, and we therefore need to know the position and length of each element. First, the level of the element is found in MATLAB by j=floor(log2(k)) + 1. As an example, the elements enumerated as 4, 5, 6, and 7 are on the third level, and the integer parts of $\log_2$ of these are 2, so the formula yields j=3. Note that the levels start at 1 and not 0 (as prescribed by the theory). This is solely due to MATLAB's inability to use index 0. Thus the reader should keep this change in numbering of the levels in mind, when referring back to the theory.

Now we find the length of an element on the j'th level as (with $j = 1$ being the first level)

$$\frac{\text{length of signal}}{\text{number of elements on the } j\text{'th level}} = \frac{2^N}{2^{j-1}} = 2^{N-j+1} .$$

The vector, which is the first element on the j'th level is then found by D(j,1:L), where L=2^(N-j+1). The second element is D(j,1+L:2*L), and the third is D(j,1+2*L:3*L), an so on. Generally, the m'th element at level j is D(j,1+(m-1)*L:m*L).

Function 8.1 Generating a Vector with Cost Values

```
J = size(D, 1);                    % Levels in the decomp

SignalLength = size(D, 2);         % Length of signal
N = log2(SignalLength);

CostValues = zeros(1, 2^J - 1);    % Initialize cost value vector

% Apply the cost function to each element in the decomposition
for k=1:2^J-1
  j = floor(log2(k)) + 1;          % Find the level
  L = 2^(N-j+1);                    % Find length of element

  % Go through all elements on the j'th level
  for m=1:2^(j-1)
    E = D(j, 1 + (m-1)*L: m*L);     % Extract element
    CostValues(k) = CostFunc(E);    % Calculate cost value
  end
end
```

When D is a decomposition matrix, the Function 8.1 will create a vector CostValues, which contains the cost value for each element in the decomposition. The reference CostFunc is to the given cost function, which also has to be implemented (see Sect. 8.3 and Sect. 11.9).

11.8.3 Best Basis Search

The next step is to find the best basis. We let a vector Basis of the same length as CostValues represent the basis. The indexing of the two vectors is the same, that is the elements are enumerated as in Fig. 11.3.

In Fig. 11.4 we have marked a basis with shaded boxes. This basis is then represented by the vector

$$\text{Basis} = [0\ 0\ 1\ 1\ 0\ 0\ 0\ 0\ 0\ 1\ 1\ 0\ 0\ 0\ 0] .$$

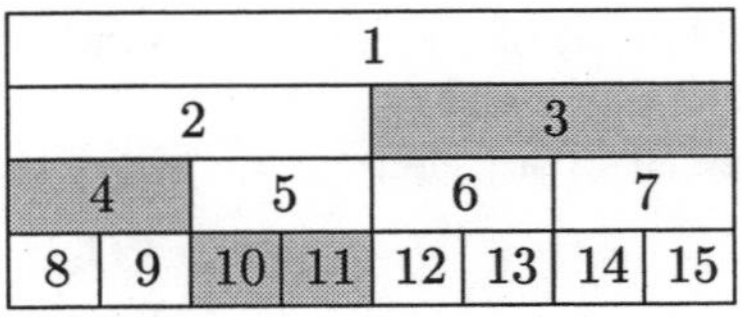

1							
2				3			
4		5		6		7	
8	9	10	11	12	13	14	15

Fig. 11.4. Basis enumeration and basis representation

In this vector we have put a 1 in the places corresponding to marked elements, and 0 elsewhere.

The implementation of the best basis algorithm on p. 94 is not difficult. First all the bottom elements are chosen, and a bottom-up search is performed. The only tricky thing is step 4(a), which requires that all marks below the marked element are deleted. Doing this would require some sort of search for 1's, and this could easily become a cumbersome procedure. Instead we temporarily leave the marks, and once the best basis search is completed, we know that the best basis is given as all the top-most 1's. To remove all the unwanted 1's we go through the binary tree again, this time top-down. Each time we encounter a 1 or a 2 in an element, we put a 2 in both elements just below it. Hence, the tree is filled up with 2 below the chosen basis, and thus removing all the unwanted 1's. Finally, all the 2's are converted to 0's.

Function 8.2 Best Basis Search
___

```
% Mark all the bottom elements.
Basis = [zeros(1, 2^(J-1)-1) ones(1, 2^(J-1))];

% Bottom-up search for the best basis.
for j=J-1:-1:1
  for k=2^(j-1):2^j-1
    v1 = CostValues(k);
    v2 = CostValues(2*k) + CostValues(2*k+1);
    if v1 >= v2
      Basis(k) = 1;
    else
      CostValues(k) = v2;
    end
  end
end

% Fill with 2's below the chosen basis.
for k=1:(length(Basis)-1)/2
  if Basis(k) == 1 | Basis(k) == 2
    Basis(2*k) = 2;
    Basis(2*k+1) = 2;
  end
end
```

```
% Convert all the 2's to 0's.
Basis = Basis .* (Basis == 1);
```

11.8.4 Other Basis Functions

Most of the properties related to the concept of a basis can be implemented in MATLAB. We have seen how to implement calculation of cost values and search for the best basis. Other useful properties which can be implemented are

- a best level search,
- displaying a basis graphically,
- checking the validity of a basis representation,
- displaying the corresponding time-frequency plane,
- reconstruction of a signal from a given basis,
- alteration of signal given by a certain basis.

The last property is important, since it is the basis for many applications, including denoising and compression. We do not present implementations of these properties, but some are discussed in the exercises.

11.9 Cost Functions

Most cost functions are easily implemented in MATLAB, since they are just functions mapping a vector to a number. For example, the ℓ^p norm is calculated as $\texttt{sum(abs(a).\^{}p)\^{}(1/p)}$, where $\texttt{a}$ is a vector (an element from the decomposition), and $\texttt{p}$ is a number between 0 and ∞.

The Shannon entropy can be a problem, however, since it involves the logarithm to the entries in $\texttt{a}$. If some of these are 0, the logarithm is undefined. We therefore have to disregard the 0 entries. This is done by $\texttt{a(find(a))}$, because $\texttt{find(a)}$ in itself returns the indices of non-zero entries. Hence, the Shannon entropy is calculated as

$$-\texttt{sum(a(find(a)).\^{}2 .* log(a(find(a)).\^{}2))}$$

Note that $\texttt{log}$ in MATLAB is the natural logarithm.

Exercises

11.1 Modify $\texttt{dwthaar}$ such that the transform becomes energy preserving (see (3.28)–(3.31)), i.e. such that $\texttt{norm(Signal) = norm(dwthaar(Signal))}$

11.2 Construct from `dwthaar` (Function 2.1) another function `idwthaar`, which implements the inverse Haar transform:

1. Construct the function such that it takes the bottom row of the decomposition as input and gives the original signal (upper most row in the decomposition) as output.
2. Construct the function such that it takes any row and the vertical location of this row in the decomposition as inputs, and gives the original signal as output.
3. Modify the function in 2. such that it becomes energy preserving.

11.3 Implement the inverse of Function 3.4, and verify that it actually computes the inverse by applying it to the output of Function 3.4 on known vectors.

11.4 One possible implementation of the inverse of the Daubechies 4 transform is shown in Function 4.5. The implementation inversely transforms the signal 'backwards' by starting at index $N/2$ instead of index 1. Implement the inverse Daubechies 4 transform, using periodization, such that it starts at index 1, and test that the implementation is correct by using Function 4.4. Do not shift the signal, just use signal samples from the other end of the signal whenever necessary.

Hint: Start all over by using the approach described at the beginning of Sect. 11.4.1 on p. 161, and the Functions 4.1, 4.2, and 4.3.

11.5 The CDF(3,3) transform is defined by the following equations.

$$s^{(1)}[n] = S[2n] - \frac{1}{3}S[2n-1] ,$$

$$d^{(1)}[n] = S[2n+1] - \frac{1}{8}(9s^{(1)}[n] + 3s^{(1)}[n+1]) ,$$

$$s^{(2)}[n] = s^{(1)}[n] + \frac{1}{36}(3d^{(1)}[n-1] + 16d^{(1)}[n] - 3d^{(1)}[n+1]) ,$$

$$s[n] = \frac{\sqrt{2}}{3}s^{(2)}[n] ,$$

$$d[n] = \frac{3}{\sqrt{2}}d^{(1)}[n] .$$

1. Write out the transform in matrix form.
2. Show that the corresponding filters are given by

$$\mathbf{h} = \frac{\sqrt{2}}{64}\begin{bmatrix} 3 & -9 & -7 & 45 & 45 & -7 & -9 & 3 \end{bmatrix} ,$$

$$\mathbf{g} = \frac{\sqrt{2}}{8}\begin{bmatrix} -1 & 3 & -3 & 1 \end{bmatrix} .$$

11.6 Write a function that performs a full wavelet decomposition to the maximal level permitted by the length of the signal. You can start from Function 2.2 on p. 155.

1. Start with a function which generates the decomposition in Table 3.1, i.e. a signal of length 8 transformed using the Haar transform.
2. Extend your function to accept signals of length 2^N.
3. Change the function to use CDF(3,3), see Exer. 11.5. Here you have to solve the boundary problem. The easy choice is zero padding.

11.7 An in-place implementation of a wavelet transform is more complicated to realize. Start by solving the following problems.

1. Explain how the entries in a transformed vector are placed, as you go through the full decomposition three. See Table 3.2 for an example.
2. Write a function which computes the location in the decomposition matrix used previously, based on the indexing used in the decomposition.
3. Write a function which implements an in place Haar transform, over a prescribed number of levels.

11.8 Implement the inverse of the wavelet transform that uses the Gram-Schmidt boundary filters in such a way that it applies to a transformed signal with separated low and high pass parts. Remember that the low and high pass boundary filters must be separated to do this.

11.9 Write a function which implements the best level basis search in a full wavelet decomposition to a prescribed level.

11.10 Not all vectors containing 0 and 1 entries can be valid representations of a basis.

1. Describe how the validity of a given vector can be checked (you have to check both length and location of 0 and 1 entries).
2. Write a function, which performs this validity check.

12. Lifting and Filters II

There are basically three forms for representing the building block in a DWT: The transform can be represented by a pair of filters (usually low pass and high pass filters) satisfying the perfect reconstruction conditions from Chap. 7, or it can be given as lifting steps, which are either given in the time domain as a set of equations, or in the frequency domain as a factored matrix of Laurent polynomials. The Daubechies 4 transform has been presented in all three forms in previous chapters, but so far we have only made casual attempts to convert between the various representations. When trying to do so, it turns out that only one conversion requires real work, namely conversion from filter to matrix and equation forms. In Chap. 7 we presented the theorem, which shows that it is always possible to do this conversion, but we did not show how to do it. This chapter is therefore dedicated to discussing the three basic forms of representation of the wavelet transform, as well as the conversions between them. In particular, we give a detailed proof of the 'from filter to matrix/equation' theorem stated in Chap. 7. The proof is a detailed and exemplified version of the proof found in I. Daubechies and W. Sweldens [7].

12.1 The Three Basic Representations

We begin by reviewing the three forms of representation, using the Daubechies 4 transform as an example.

Matrix form:

$$\mathbf{H}(z) = \begin{bmatrix} \frac{\sqrt{3}-1}{\sqrt{2}} & 0 \\ 0 & \frac{\sqrt{3}+1}{\sqrt{2}} \end{bmatrix} \begin{bmatrix} 1 & -z \\ 0 & 1 \end{bmatrix} \begin{bmatrix} 1 & 0 \\ -\frac{\sqrt{3}}{4} - \frac{\sqrt{3}-2}{4}z^{-1} & 1 \end{bmatrix} \begin{bmatrix} 1 & \sqrt{3} \\ 0 & 1 \end{bmatrix}. \quad (12.1)$$

Equation form:

$$s^{(1)}[n] = S[2n] + \sqrt{3}S[2n+1], \quad (12.2)$$

$$d^{(1)}[n] = S[2n+1] - \frac{1}{4}\sqrt{3}s^{(1)}[n] - \frac{1}{4}(\sqrt{3}-2)s^{(1)}[n-1], \quad (12.3)$$

$$s^{(2)}[n] = s^{(1)}[n] - d^{(1)}[n+1], \quad (12.4)$$

$$\tilde{s}[n] = \frac{\sqrt{3}-1}{\sqrt{2}}s^{(2)}[n] \, , \tag{12.5}$$

$$\tilde{d}[n] = \frac{\sqrt{3}+1}{\sqrt{2}}d^{(1)}[n] \, . \tag{12.6}$$

Filter form:

$$\mathbf{h} = \frac{1}{4\sqrt{2}}\left[1+\sqrt{3}, \, 3+\sqrt{3}, \, 3-\sqrt{3}, \, 1-\sqrt{3}\right] \, , \tag{12.7}$$

$$\mathbf{g} = \frac{1}{4\sqrt{2}}\left[1-\sqrt{3}, \, -3+\sqrt{3}, \, 3+\sqrt{3}, \, -1-\sqrt{3}\right] \, . \tag{12.8}$$

Depending on the circumstances each form has its advantages. In an implementation it is always either the equation form (when implementing as lifting steps) or the filter form (when implementing as a filter bank) which is used. However, when concerned with the theoretical aspects of the lifting theory, the matrix form is very useful. Moreover, if we want to design a filter via lifting steps (some basic steps for this were presented in Sect. 3.2 and 3.3), but use it in a filter bank, we need a tool for converting the lifting steps, in either matrix or equation form, to the filter form. On the other hand, if we want to use existing filters as lifting steps, we need to convert the filter form to the equation form. In brief, it is very useful to be able to convert between the three forms. Here is a list of where we present the various conversions.

Matrix ↔ equation	Sect. 12.2
Matrix → filter	Sect. 7.3
Equation → filter	Sect. 12.3
Filter → matrix	Sect. 12.4
Filter → equation	Sect. 12.4

The only real challenge is converting the filter form to the matrix form, or the equation form. But first we do the easy conversions.

12.2 From Matrix to Equation Form

The factored matrix form (12.1) is closely related to the equation form. Each 2×2 matrix corresponding to one equation, except for the first matrix, which corresponds to the two scale equations. Note that due to the way matrix multiplication is defined, the steps appear in the reverse order in the matrix form. The last matrix is the first one to be applied to the two components of the signal, and the first matrix in the product is the normalization step.

When using the matrix form for transforming a signal, the signal is given by its z-transform $S(z)$, which is split into its even and odd components and placed in a vector. So

$$\mathbf{H}(z)\begin{bmatrix} S_0(z) \\ S_1(z) \end{bmatrix} = \begin{bmatrix} S_{\mathrm{L}}(z) \\ S_{\mathrm{H}}(z) \end{bmatrix},$$

where $S_{\mathrm{L}}(z)$ and $S_{\mathrm{H}}(z)$ denote the z-transform of the low and high pass transformed signals respectively. We now apply $\mathbf{H}(z)$ one matrix at a time.

$$\begin{bmatrix} 1 & \sqrt{3} \\ 0 & 1 \end{bmatrix}\begin{bmatrix} S_0(z) \\ S_1(z) \end{bmatrix} = \begin{bmatrix} S_0(z) + \sqrt{3}S_1(z) \\ S_1(z) \end{bmatrix} \equiv \begin{bmatrix} S^{(1)}(z) \\ S_1(z) \end{bmatrix},$$

where $S^{(1)}(z)$ is the intermediate step notation for $S_0(z) + \sqrt{3}S_1(z)$. Multiplication with the second matrix gives

$$\begin{bmatrix} 1 & 0 \\ -\frac{\sqrt{3}}{4} - \frac{\sqrt{3}-2}{4}z^{-1} & 1 \end{bmatrix}\begin{bmatrix} S^{(1)}(z) \\ S_1(z) \end{bmatrix}$$

$$= \begin{bmatrix} S^{(1)}(z) \\ S_1(z) - \frac{\sqrt{3}}{4}S^{(1)}(z) - \frac{\sqrt{3}-2}{4}z^{-1}S^{(1)}(z) \end{bmatrix} \equiv \begin{bmatrix} S^{(1)}(z) \\ D^{(1)}(z) \end{bmatrix},$$

while multiplication with the third matrix gives

$$\begin{bmatrix} 1 & -z \\ 0 & 1 \end{bmatrix}\begin{bmatrix} S^{(1)}(z) \\ D^{(1)}(z) \end{bmatrix} = \begin{bmatrix} S^{(1)}(z) - zD^{(1)}(z) \\ D^{(1)}(z) \end{bmatrix} \equiv \begin{bmatrix} S^{(2)}(z) \\ D^{(1)}(z) \end{bmatrix}.$$

Finally the scaling matrix is applied. Collecting all the three equations we have

$$S^{(1)}(z) = S_0(z) + \sqrt{3}S_1(z),$$

$$D^{(1)}(z) = S_1(z) - \frac{\sqrt{3}}{4}S^{(1)}(z) - \frac{\sqrt{3}-2}{4}z^{-1}S^{(1)}(z),$$

$$S^{(2)}(z) = S^{(1)}(z) - zD^{(1)}(z),$$

$$\tilde{S}(z) = \frac{\sqrt{3}-1}{\sqrt{2}}S^{(2)}(z),$$

$$\tilde{D}(z) = \frac{\sqrt{3}+1}{\sqrt{2}}D^{(1)}(z).$$

To get the equation form we use the definition of the z-transform and the uniqueness of the z-transform representation. The original signal is in the time-domain given by the sequence $\{S[n]\}$. Thus

$$S_0(z) = \sum_n S[2n]z^{-n} \quad \text{and} \quad S_1(z) = \sum_n S[2n+1]z^{-n},$$

see also (7.16)–(7.25). Using the notation $S^{(1)}(z) = \sum_n s^{(1)}[n]z^{-n}$, as in Chap. 7, we have that the first equation reads

$$\sum_n s^{(1)}[n]z^{-n} = \sum_n (s[2n] + \sqrt{3}s[2n+1])z^{-n},$$

and then uniqueness of the z-transform representation yields (12.2). The remaining equations follow by similar computations, if we also use the shifting properties associated with multiplication by z, see (7.10), and by z^{-1}, see (7.11).

The other way, from equation to matrix form, is done by the reverse procedure (see Exer. 12.4).

12.3 From Equation to Filter Form

In the previous chapters we have repeatedly seen how references back and forth in time in the equation form can be handle by inserting appropriate expressions instead of the reference. Actually, this can be done for all references. We do this systematically by starting with the last equation and work our way up.

This is here exemplified with Daubechies 4. We begin with the last equation (12.6)

$$\tilde{d}[n] = \frac{\sqrt{3}+1}{\sqrt{2}} d^{(1)}[n] \ .$$

The reference to $d^{(1)}[n]$ can be replaced by the actual expression for $d^{(1)}[n]$.

$$\tilde{d}[n] = \frac{\sqrt{3}+1}{\sqrt{2}} \left(s[2n+1] - \frac{\sqrt{3}}{4} s^{(1)}[n] - \frac{\sqrt{3}-2}{4} s^{(1)}[n-1] \right) \ .$$

Then we insert $s^{(1)}[n]$ and $s^{(1)}[n-1]$.

$$\begin{aligned}
\tilde{d}[n] &= \frac{\sqrt{3}+1}{\sqrt{2}} \left(s[2n+1] - \frac{\sqrt{3}}{4} \left(s[2n] + \sqrt{3}s[2n+1] \right) \right. \\
&\qquad\qquad \left. - \frac{\sqrt{3}-2}{4} \left(s[2n-2] + \sqrt{3}s[2n-1] \right) \right) \\
&= \frac{\sqrt{3}-1}{4\sqrt{2}} s[2n-2] + \frac{3-\sqrt{3}}{4\sqrt{2}} s[2n-1] - \frac{\sqrt{3}+3}{4\sqrt{2}} s[2n] + \frac{\sqrt{3}+1}{4\sqrt{2}} s[2n+1] \\
&= \sum_{m=-1}^{2} g[m] s[2n-m] \ ,
\end{aligned}$$

where **g** is the high pass impulse response (12.8) of Daubechies 4 (except for a change of sign). In the same manner we find

$$\begin{aligned}
\tilde{s}[n] &= \frac{1+\sqrt{3}}{4\sqrt{2}} s[2n-2] + \frac{3+\sqrt{3}}{4\sqrt{2}} s[2n-1] + \frac{3-\sqrt{3}}{4\sqrt{2}} s[2n] + \frac{1-\sqrt{3}}{4\sqrt{2}} s[2n+1] \\
&= \sum_{m=-1}^{2} h[m] s[2n-m] \ ,
\end{aligned}$$

where **h** is the low pass impulse response (12.7) of Daubechies 4. The rewritten expressions for $\tilde{s}[n]$ and $\tilde{d}[n]$ show that they correspond to a convolution of the impulse responses and four samples. As n takes the values from 0 to half the length of the original signal, a filtering has occurred.

There are two distinct differences between the filter form and the equation form. Filtering the signal requires approximately twice as many calculations compared to 'lifting it' (see Exer. 12.1), but the filter form is much more easily implemented, since the convolution is independent of the structure of the signal and of the transform. It is unfortunately not possible to have at the same time both efficient and general implementations of the equation form.

12.4 From Filters to Lifting Steps

We have now seen how to make lifting steps into ordinary FIR filters. This presented no particular challenge, since it was merely a matter of expanding equations. The other direction, i.e. making filters into lifting steps, is a bit more tricky, since we now have to factorize the polyphase matrix. In Chap. 7 we showed with Theorem 7.3.1 that it could be done, but we omitted the constructive proof. This section is therefore dedicated to a thorough discussion of the proof, whereby the algorithm for factoring the polyphase matrix is given. The proof given here is mainly due to I. Daubechies and W. Sweldens [7]. First, we restate the theorem.

Theorem 12.4.1. *Given a 2×2 matrix*

$$\mathbf{H}(z) = \begin{bmatrix} H_{00}(z) & H_{01}(z) \\ H_{10}(z) & H_{11}(z) \end{bmatrix},$$

where the $H_{nk}(z)$ are Laurent polynomials, and where

$$\det \mathbf{H}(z) = H_{00}(z)H_{11}(z) - H_{01}(z)H_{10}(z) = 1, \qquad (12.9)$$

then there exist a constant $K \neq 0$ and Laurent polynomials

$$S_1(z), \dots, S_M(z), \quad and \quad T_1(z), \dots, T_M(z),$$

such that

$$\mathbf{H}(z) = \begin{bmatrix} K & 0 \\ 0 & K^{-1} \end{bmatrix} \prod_{k=1}^{M} \begin{bmatrix} 1 & S_k(z) \\ 0 & 1 \end{bmatrix} \begin{bmatrix} 1 & 0 \\ T_k(z) & 1 \end{bmatrix}. \qquad (12.10)$$

This theorem requires the matrix to have determinant 1, so in order to apply it to an analysis wavelet filter pair, we need H_{nk} of the filter to fulfill (12.9). The question is therefore whether this is always the case for wavelet filters.

We know that $\mathbf{H}(z)$ performs the DWT of a signal in the z-transform, and we know that this transform is invertible. Therefore $\mathbf{H}^{-1}(z)$ exists, and according to Proposition 7.2.1 $\det \mathbf{H}(z) = a^{-1}z^{-n}$ for some $a \neq 0$ and n. Consequently,

$$\det \begin{bmatrix} H_{00}(z) & H_{01}(z) \\ az^n H_{10}(z) & az^n H_{11}(z) \end{bmatrix} = 1 \ .$$

So the determinant 1 requirement can always be fulfilled with a wavelet filter by choosing the proper z-transform.

We are now ready to begin the proof of Theorem 12.4.1. It is presented in the following sections along with some examples of factorization. Note that we now use the same notation as in Chap. 7, that is $H_0(z)$ for the low pass analysis filter and $H_1(z)$ for the high pass analysis filter, and $G_0(z)$ and $G_1(z)$ for the synthesis filters.

12.4.1 The Euclidean Algorithm

The Euclidean algorithm is usually first presented as an algorithm for finding the greatest common divisor of two integers. But it can be applied to many other analogous problems. One application is to finding the greatest common divisor of two polynomials. Here we apply it to Laurent polynomials. We recall that the z-transform of a FIR filter $\mathbf{h}$ is a Laurent polynomial, which is a polynomial of the form

$$h(z) = \sum_{k=k_b}^{k_e} h[k]z^{-k} \ .$$

Here k_b and k_e are integers satisfying $k_b \leq k_e$. This is in contrast to ordinary polynomials, where we only have nonnegative powers. To define the degree of a Laurent polynomial, we assume that $h[k_b] \neq 0$ and $h[k_e] \neq 0$. Then the *degree* of $h(z)$ is defined as

$$|h| = k_e - k_b \ .$$

The zero Laurent polynomial is assigned the degree $-\infty$. A polynomial of degree 0, such as $3z^7$, is also referred to as a monomial.

Take two Laurent polynomials $a(z)$ and $b(z) \neq 0$ with $|a(z)| \geq |b(z)|$. Then there always exist a Laurent polynomial $q(z)$, the quotient, with $|q(z)| = |a(z)| - |b(z)|$, and a Laurent polynomial $r(z)$, the remainder, with $|r(z)| < |b(z)|$ such that

$$a(z) = b(z)q(z) + r(z) \ . \tag{12.11}$$

We use the notation

$$q(z) = a(z)/b(z), \quad \text{and} \quad r(z) = a(z)\%b(z) \ .$$

If $b(z)$ is a monomial, then $r(z) = 0$, and the division is exact. A Laurent polynomial is invertible, if and only if it is a monomial, see the proof of Proposition 7.2.1. In other words, the only Laurent polynomials with product equal to 1 are pairs αz^m and $\alpha^{-1} z^{-m}$.

The division (12.11) can be repeated with $b(z)$ and $r(z)$, as in

$$b(z) = r(z)q_1(z) + r_1(z) . \tag{12.12}$$

Since the degree of the remainder decreases by at least one, it takes at most $|b(z)| + 1$ steps to achieve a remainder equaling 0. This argument proves the following theorem.

Theorem 12.4.2 (Euclidean Algorithm for Laurent Polynomials). *Given two Laurent polynomials $a(z)$ and $b(z) \neq 0$, such that $|a(z)| \geq |b(z)|$. Let $a_0(z) = a(z)$ and $b_0(z) = b(z)$, and iterate the following steps starting from $n = 0$*

$$a_{n+1}(z) = b_n(z) , \tag{12.13}$$
$$b_{n+1}(z) = a_n(z) \% b_n(z) . \tag{12.14}$$

Let N denote the smallest integer with $N \leq |b(z)| + 1$, for which $b_N(z) = 0$. Then $a_N(z)$ is a greatest common divisor for $a(z)$ and $b(z)$.

We note that there is no unique greatest common divisor for $a(z)$ and $b(z)$, since if $d(z)$ divides both $a(z)$ and $b(z)$, and is of maximal degree, then $\alpha z^m d(z)$ is also a divisor of the same degree.

This theorem is the key to constructing the 2×2 matrix lifting steps, since each iteration will produce one matrix, as we show below.

It is important to note that $q(z)$ and $r(z)$ in (12.11) are not unique. Usually there is more than one valid choice. This is easily seen with an example. Let

$$a(z) = -z^{-1} + 6 - z .$$
$$b(z) = 2z^{-1} + 2 .$$

Then $q(z)$ is necessarily on the form $c + dz$, so from (12.11)

$$-z^{-1} + 6 - z = 2cz^{-1} + 2c + 2d + 2dz + r(z) . \tag{12.15}$$

By proper choice of c and d we can match at least two of the three terms (if we could match all three terms we would have an exact division and thus $r(z) = 0$). Let us match the two first in (12.15), that is terms z^{-1} and z^0. Then

$$c = -\frac{1}{2} \quad \text{and} \quad 2\left(-\frac{1}{2} + d\right) = 6 \quad \Leftrightarrow \quad d = \frac{7}{2} .$$

Since $r(z) = a(z) - b(z)q(z)$ we find that

$$r(z) = -z^{-1} + 6 - z - (-z^{-1} + 6 + 7z) = -8z .$$

If we instead first match the two z terms in (12.15) we get $d = -1/2$, and if we then match the two z^{-1}, we get $c = -1/2$. Thus

$$r(z) = -z^{-1} + 6 - z - (-z^{-1} - 2 - z) = 8 .$$

Both factorizations of $a(z)$ are valid, and both will serve the purpose of Theorem 12.4.2.

12.4.2 The Euclidean Algorithm in Matrix Form

The first step in proving Theorem 12.4.1 is to examine the iterations defined in Theorem 12.4.2 and rewrite them in the form of a product of 2×2 matrices, whose entries are Laurent polynomials. If we let $q_{n+1}(z) = a_n(z)/b_n(z)$, then the first step in the algorithm is

$$a_1(z) = b_0(z) ,$$
$$b_1(z) = a_0(z) - b_0(z)q_1(z) ,$$

the next steps is

$$a_2(z) = b_1(z) ,$$
$$b_2(z) = a_1(z) - b_1(z)q_2(z) ,$$

and after N steps

$$a_N(z) = b_{N-1}(z) ,$$
$$0 = b_N(z) = a_{N-1}(z) - b_{N-1}(z)q_N(z) .$$

Note that according to the theorem $b_N(z) = 0$. The first step can also be written

$$\begin{bmatrix} a_1(z) \\ b_1(z) \end{bmatrix} = \begin{bmatrix} 0 & 1 \\ 1 & -q_1(z) \end{bmatrix} \begin{bmatrix} a(z) \\ b(z) \end{bmatrix} ,$$

while the second step becomes

$$\begin{bmatrix} a_2(z) \\ b_2(z) \end{bmatrix} = \begin{bmatrix} 0 & 1 \\ 1 & -q_2(z) \end{bmatrix} \begin{bmatrix} a_1(z) \\ b_1(z) \end{bmatrix} = \begin{bmatrix} 0 & 1 \\ 1 & -q_2(z) \end{bmatrix} \begin{bmatrix} 0 & 1 \\ 1 & -q_1(z) \end{bmatrix} \begin{bmatrix} a(z) \\ b(z) \end{bmatrix} .$$

Finally we get

$$\begin{bmatrix} a_N(z) \\ 0 \end{bmatrix} = \prod_{n=N}^{1} \begin{bmatrix} 0 & 1 \\ 1 & -q_n(z) \end{bmatrix} \begin{bmatrix} a(z) \\ b(z) \end{bmatrix} .$$

Note the order of the terms in this product, as given by the limits in the product. The term with index N is at the left end, and the one with index 1 is at the right end. We now note that the inverse of

$$\begin{bmatrix} 0 & 1 \\ 1 & -q_n(z) \end{bmatrix} \quad \text{is} \quad \begin{bmatrix} q_n(z) & 1 \\ 1 & 0 \end{bmatrix},$$

as the reader immediately can verify. Thus we can multiply by these inverse matrices on the in the equation. Consequently

$$\begin{bmatrix} a(z) \\ b(z) \end{bmatrix} = \prod_{n=1}^{N} \begin{bmatrix} q_n(z) & 1 \\ 1 & 0 \end{bmatrix} \begin{bmatrix} a_N(z) \\ 0 \end{bmatrix}. \tag{12.16}$$

Let us now apply this to the low pass filter $H_0(z)$. The polyphase components of $H_0(z)$ are $H_{00}(z)$ and $H_{01}(z)$, and if we let $a(z) = H_{00}(z)$ and $b(z) = H_{01}(z)$ we get

$$\begin{bmatrix} H_{00}(z) \\ H_{01}(z) \end{bmatrix} = \prod_{n=1}^{N} \begin{bmatrix} q_n(z) & 1 \\ 1 & 0 \end{bmatrix} \begin{bmatrix} Kz^c \\ 0 \end{bmatrix}. \tag{12.17}$$

Notice that we get a monomial as a greatest common divisor of $H_{00}(z)$ and $H_{01}(z)$. This can be seen as follows. We know that

$$H_{00}(z)H_{11}(z) - H_{01}(z)H_{10}(z) = 1. \tag{12.18}$$

Let $p(z)$ denote a greatest common divisor of $H_{00}(z)$ and $H_{01}(z)$. This means that both $H_{00}(z)$ and $H_{01}(z)$ can be divided by $p(z)$, and hence the entire left hand side in (12.18) can be divided by $p(z)$. But then the right hand side must also be divisible by $p(z)$, and the only Laurent polynomial that divides 1 is a monomial. Hence $p(z) = Kz^c$. Since the theorem stated that $a_N(z)$ was one of the greatest common divisors, and since common divisors differ only by a monomial factor, we then deduce that $a_N(z)$ is a monomial.

If we moreover multiply (from the right) on both sides of (12.17) with z^{-c} we get

$$\begin{bmatrix} H'_{00}(z) \\ H'_{01}(z) \end{bmatrix} = \begin{bmatrix} z^{-c}H_{00}(z) \\ z^{-c}H_{01}(z) \end{bmatrix} = \prod_{n=1}^{N} \begin{bmatrix} q_n(z) & 1 \\ 1 & 0 \end{bmatrix} \begin{bmatrix} K \\ 0 \end{bmatrix}. \tag{12.19}$$

Multiplying $H_{00}(z)$ with z^{-c} only shifts the indices in the z-transform, and hence does not change the fact that it is the even coefficients of the low pass impulse response. In other words by choosing the right z-transformation of the low pass impulse response, it is always possible to end up with $a_N(z) = K$.

12.4.3 Example on Factoring a Laurent Polynomial

Before we continue with the proof of the factorization theorem, let us clarify the above results with an example. We use CDF(2,2) to show how the factorization works. The low pass impulse response is given by

$$\frac{\sqrt{2}}{8} \left[-1\ 2\ 6\ 2\ -1 \right] .$$

We begin with the symmetric transform (omitting the scaling $\sqrt{2}/8$)

$$H_0(z) = -z^{-2} + 2z^{-1} + 6 + 2z - z^2 . \qquad (12.20)$$

The polyphase components are then

$$H_{00}(z) = -z^{-1} + 6 - z, \quad \text{and} \quad H_{01}(z) = 2z^{-1} + 2 ,$$

according to (7.38) on p. 72. The first step in the algorithm, that is theorem 12.4.2, is

$$a_0(z) = -z^{-1} + 6 - z ,$$
$$b_0(z) = 2z^{-1} + 2 .$$

If we match terms from the left (see description at the end of Sect. 12.4.1), we get

$$-z^{-1} + 6 - z = (2z^{-1} + 2) \cdot \left(-\frac{1}{2} + \frac{7}{2}z \right) - 8z , \qquad (12.21)$$

such that

$$q_1(z) = -\frac{1}{2} + \frac{7}{2}z ,$$
$$r_1(z) = -8z .$$

The next steps is then

$$a_1(z) = b_0(z) = 2z^{-1} + 2 ,$$
$$b_1(z) = r_1(z) = -8z .$$

Again matching from the left (although is does not matter in this particular case)

$$2z^{-1} + 2 = (-8z) \cdot \left(-\frac{1}{4}z^{-2} - \frac{1}{4}z^{-1} \right) ,$$

such that

$$q_2(z) = -\frac{1}{4}z^{-2} - \frac{1}{4}z^{-1} ,$$
$$r_2(z) = 0 .$$

Finally

$$a_2(z) = b_1(z) = -8z ,$$
$$b_2(z) = r_2(z) = 0 .$$

Since $b_2(z) = 0$, we have found a greatest common divisor of $a_0(z)$ and $b_0(z)$, namely $a_2(z) = -8z$. Putting all this into (12.17) yields

$$
\begin{bmatrix} H_{00}(z) \\ H_{01}(z) \end{bmatrix} = \begin{bmatrix} -z^{-1} + 6 - z \\ 2z^{-1} + 2 \end{bmatrix}
$$

$$
= \begin{bmatrix} -\frac{1}{2} + \frac{7}{2}z & 1 \\ 1 & 0 \end{bmatrix} \begin{bmatrix} -\frac{1}{4}z^{-2} - \frac{1}{4}z^{-1} & 1 \\ 1 & 0 \end{bmatrix} \begin{bmatrix} -8z \\ 0 \end{bmatrix}. \tag{12.22}
$$

Unfortunately we did not get a constant in the last vector. This can be achieved through multiplication by z^{-1} on both sides

$$
\begin{bmatrix} -z^{-2} + 6z^{-1} - 1 \\ 2z^{-2} + 2z^{-1} \end{bmatrix} = \begin{bmatrix} -\frac{1}{2} + \frac{7}{2}z & 1 \\ 1 & 0 \end{bmatrix} \begin{bmatrix} -\frac{1}{4}z^{-2} - \frac{1}{4}z^{-1} & 1 \\ 1 & 0 \end{bmatrix} \begin{bmatrix} -8 \\ 0 \end{bmatrix}. \tag{12.23}
$$

So if we had chosen the z-transform

$$
H_0(z) = -z^{-4} + 2z^{-3} + 6z^{-2} + 2z^{-1} - 1 ,
$$

instead of (12.20) the gcd would have been a constant. Note that choosing a factorization that does not give a constant gcd is by no means fatal. In fact, no matter what z-transform we start with, we get the same matrices (provide that the same matching of terms is used), and the only difference is the gcd. So if the gcd is not constant, simply keep the coefficient and discard whatever power of z is present. This is exactly what we just did; the only difference between the lifting steps (the right hand sides) in (12.22) and (12.23) is that the z in the former is discarded in the latter.

But this factorization is not our only option. If we had chosen another matching in the first step, say

$$
-z^{-1} + 6 - z = (2z^{-1} + 2) \cdot \left(-\frac{1}{2} - \frac{1}{2}z \right) + 8 ,
$$

instead of (12.21) we would end up with

$$
\begin{bmatrix} -z^{-1} + 6 - 1z \\ 2z^{-1} + 2 \end{bmatrix} = \begin{bmatrix} -\frac{1}{2} - \frac{1}{2}z & 1 \\ 1 & 0 \end{bmatrix} \begin{bmatrix} \frac{1}{4}z^{-1} + \frac{1}{4} & 1 \\ 1 & 0 \end{bmatrix} \begin{bmatrix} 8 \\ 0 \end{bmatrix},
$$

which does not need any modification. Incidentally, this is the form of the equation (3.36) and (3.37) on p. 23, and if we multiply 8 with $\sqrt{2}/8$, the omitted scaling of the impulse response, we get $\sqrt{2}$, the scaling of the low pass part in (3.41).

The important point to note here is that since division with remainder of Laurent polynomials is not unique, neither is the factorization into lifting steps. The fact that the gcd in some cases is not a constant is a trivial problem, as described above. But a more serious, and definitely not trivial, problem arises. While the first factorization had a factor 7/2, the second factorization had no factors larger than 1/2 (disregarding the final scaling factor). This means that although the output of the complete transform is never larger than 2 times the input, intermediate calculations has the potential of becoming at least 3.5 times larger than the input. This may not seem to be a serious problem. However, for longer filters the values in the intermediate calculations can becomes significantly larger. In Sect. 12.5 we give another example which demonstrates this phenomenon. Stated briefly, it is important to choose the right factorization.

12.4.4 Completing the Factorization

We still have some more work to do before the final factorization is achieved, and from now on we assume that the factorization is done such that $a_N(z)$ is a constant. The form of (12.19) is not entirely the same as the form of (12.10). It can be made a little more alike if we observe that

$$\begin{bmatrix} q(z) & 1 \\ 1 & 0 \end{bmatrix} = \begin{bmatrix} 1 & q(z) \\ 0 & 1 \end{bmatrix} \begin{bmatrix} 0 & 1 \\ 1 & 0 \end{bmatrix} = \begin{bmatrix} 0 & 1 \\ 1 & 0 \end{bmatrix} \begin{bmatrix} 1 & 0 \\ q(z) & 1 \end{bmatrix}.$$

Using the first equation for odd n, and the second one for even n, gives

$$\begin{bmatrix} H_{00}(z) \\ H_{01}(z) \end{bmatrix} = \prod_{n=1}^{N/2} \begin{bmatrix} 1 & q_{2n-1}(z) \\ 0 & 1 \end{bmatrix} \begin{bmatrix} 1 & 0 \\ q_{2n}(z) & 1 \end{bmatrix} \begin{bmatrix} K \\ 0 \end{bmatrix}. \tag{12.24}$$

If N is odd, we take $q_{2n}(z) = 0$. If we now replace

$$\begin{bmatrix} K \\ 0 \end{bmatrix} \quad \text{with} \quad \begin{bmatrix} K & 0 \\ 0 & K^{-1} \end{bmatrix},$$

we get

$$\begin{bmatrix} H_{00}(z) & H'_{10}(z) \\ H_{01}(z) & H'_{11}(z) \end{bmatrix} = \prod_{n=1}^{N/2} \begin{bmatrix} 1 & q_{2n-1}(z) \\ 0 & 1 \end{bmatrix} \begin{bmatrix} 1 & 0 \\ q_{2n}(z) & 1 \end{bmatrix} \begin{bmatrix} K & 0 \\ 0 & K^{-1} \end{bmatrix}, \tag{12.25}$$

where these equations *define* $H'_{10}(z)$ and $H'_{11}(z)$. By transposing both sides (remember that the transpose of a matrix product is the transpose of each matrix, multiplied in the reverse order) we get the following result, which is closer to the goal.

$$\begin{bmatrix} H_{00}(z) & H_{01}(z) \\ H'_{10}(z) & H'_{11}(z) \end{bmatrix} = \begin{bmatrix} K & 0 \\ 0 & K^{-1} \end{bmatrix} \prod_{n=M}^{1} \begin{bmatrix} 1 & q_{2n}(z) \\ 0 & 1 \end{bmatrix} \begin{bmatrix} 1 & 0 \\ q_{2n-1}(z) & 1 \end{bmatrix}. \tag{12.26}$$

All we need to do now is to find out how $H'_{10}(z)$ and $H'_{11}(z)$ are connected with $H_{10}(z)$ and $H_{11}(z)$.

To do this we observe that if the analysis filter pair $\big(H_0(z), H_1(z)\big)$ has a polyphase representation with determinant 1, then any other analysis filter pair $\big(H_0(z), H_1^{\text{new}}(z)\big)$ is related by

$$H_1^{\text{new}}(z) = H_1(z) + H_0(z)t(z^2),$$

where $t(z)$ is a Laurent polynomial. To verify this we need to show that the determinant of the polyphase matrix of $(H_0(z), H_1^{\text{new}}(z))$ is 1.

$$\mathbf{H}^{\text{new}}(z) = \begin{bmatrix} H_{00}(z) & H_{01}(z) \\ H_{10}^{\text{new}}(z) & H_{11}^{\text{new}}(z) \end{bmatrix}$$

$$= \begin{bmatrix} H_{00}(z) & H_{01}(z) \\ H_{10}(z) + H_{00}t(z) & H_{11}(z) + H_{01}(z)t(z) \end{bmatrix}$$

$$= \begin{bmatrix} 1 & 0 \\ t(z) & 1 \end{bmatrix} \begin{bmatrix} H_{00}(z) & H_{01}(z) \\ H_{10}(z) & H_{11}(z) \end{bmatrix}.$$

It follows that $\det \mathbf{H}^{\text{new}}(z) = \det \mathbf{H}(z) = 1$.

Applying this result to (12.26), we can get the original high pass filter $H_1(z)$ in the following way. From the previous calculation we know that there exists a Laurent polynomial $t(z)$ such that

$$\begin{bmatrix} H_{00}(z) & H'_{10}(z) \\ H_{01}(z) & H'_{11}(z) \end{bmatrix} \begin{bmatrix} -t(z) \\ 1 \end{bmatrix} = \begin{bmatrix} H_{10}(z) \\ H_{11}(z) \end{bmatrix},$$

and by multiplying on both side with the inverse of the 2×2 matrix, we find that

$$\begin{bmatrix} -t(z) \\ 1 \end{bmatrix} = \begin{bmatrix} H'_{11}(z) & -H'_{10}(z) \\ -H_{01}(z) & H_{00}(z) \end{bmatrix} \begin{bmatrix} H_{10}(z) \\ H_{11}(z) \end{bmatrix},$$

and thus

$$t(z) = H'_{10}(z)H_{11}(z) - H'_{11}(z)H_{10}(z). \tag{12.27}$$

Thus, multiplying (12.26) from the left with

$$\begin{bmatrix} 1 & 0 \\ -t(z) & 1 \end{bmatrix}$$

gives

$$\begin{bmatrix} H_{00}(z) & H_{01}(z) \\ H_{10}(z) & H_{11}(z) \end{bmatrix} = \begin{bmatrix} K & 0 \\ 0 & K^{-1} \end{bmatrix} \begin{bmatrix} 1 & 0 \\ -K^2 t(z) & 1 \end{bmatrix} \prod_{n=M}^{1} \begin{bmatrix} 1 & q_{2n}(z) \\ 0 & 1 \end{bmatrix} \begin{bmatrix} 1 & 0 \\ q_{2n-1}(z) & 1 \end{bmatrix}, \tag{12.28}$$

where we have used the simple relation

$$\begin{bmatrix} 1 & 0 \\ -t(z) & 1 \end{bmatrix} \begin{bmatrix} K & 0 \\ 0 & K^{-1} \end{bmatrix} = \begin{bmatrix} K & 0 \\ 0 & K^{-1} \end{bmatrix} \begin{bmatrix} 1 & 0 \\ -K^2 t(z) & 1 \end{bmatrix}.$$

By a suitable reindexing of the q polynomials (and at the same time making $K^2 t(z)$ one of them), it is now possible to determine the $S(z)$ and $T(z)$ in Theorem 12.4.1.

This concludes the constructive proof the lifting theorem. In the next sections we will give examples and show that there can be numerical problems in this constructive procedure.

12.5 Factoring Daubechies 4 into Lifting Steps

We now give two examples of creating lifting steps using the algorithm presented in the previous sections. The first example is Daubechies 4 which should be well-known by now, since we have discussed it in Sect. 3.4 and Sect. 7.3. Since we have the exact filter taps in (7.76) on p. 83, we can also find the exact lifting steps. The other example is Coiflet 12, see I. Daubechies [6], in which case the exact filter taps are not available, and we therefore have to do the calculations numerically. This second examples demonstrates not only how to handle a longer filter (which is not much different from Daubechies 4), but also the importance of choosing the right factorization.

The Daubechies 4 filter taps are given by

$$\mathbf{H}_0 = \begin{bmatrix} \frac{1+\sqrt{3}}{4\sqrt{2}} & \frac{3+\sqrt{3}}{4\sqrt{2}} & \frac{3-\sqrt{3}}{4\sqrt{2}} & \frac{1-\sqrt{3}}{4\sqrt{2}} \end{bmatrix}.$$

The even and odd coefficients are separated into

$$H_{00}(z) = a_0(z) = \frac{3+\sqrt{3}}{4\sqrt{2}} + \frac{1-\sqrt{3}}{4\sqrt{2}} z \,,$$

$$H_{01}(z) = b_0(z) = \frac{1+\sqrt{3}}{4\sqrt{2}} + \frac{3-\sqrt{3}}{4\sqrt{2}} z \,.$$

Remember that the choice of z-transform does not matter for the final factorization (see end of Sect. 12.4.3), and we choose the z-transform with no negative powers. The first step is to find $q_1(z)$. Since $a_0(z)$ and $b_0(z)$ have the same degree, the quotient is a monomial. Matching from the left yields

$$q_1(z) = \frac{\text{leftmost term of } a_0(z)}{\text{leftmost term of } b_0(z)} = \frac{\frac{3+\sqrt{3}}{4\sqrt{2}}}{\frac{1+\sqrt{3}}{4\sqrt{2}}} = \frac{3+\sqrt{3}}{1+\sqrt{3}}$$

$$= \frac{(3+\sqrt{3})(1-\sqrt{3})}{(1+\sqrt{3})(1-\sqrt{3})} = \frac{-2\sqrt{3}}{-2} = \sqrt{3} \,.$$

The remainder is then

$$r_1(z) = a_0(z) - b_0(z)q_1(z)$$

$$= \left(\frac{3+\sqrt{3}}{4\sqrt{2}} + \frac{1-\sqrt{3}}{4\sqrt{2}} z \right) - \left(\frac{1+\sqrt{3}}{4\sqrt{2}} + \frac{3-\sqrt{3}}{4\sqrt{2}} z \right) \cdot \sqrt{3}$$

$$= \frac{1-\sqrt{3}-3\sqrt{3}+3}{4\sqrt{2}} z$$

$$= \frac{1-\sqrt{3}}{\sqrt{2}} z \,.$$

This was the first iteration. The next one begins with

$$a_1(z) = b_0(z) = \frac{1+\sqrt{3}}{4\sqrt{2}} + \frac{3-\sqrt{3}}{4\sqrt{2}}z \;,$$

$$b_1(z) = r_1(z) = \frac{1-\sqrt{3}}{\sqrt{2}}z \;.$$

This time the quotient has degree 1, since $b_1(z)$ is one degree less than $a_1(z)$. More specific, $q_2(z)$ most be on the form $cz^{-1} + d$. Matching from the left means determining c first, and matching from the right means determining d first. We will do the latter. Thus d, the constant term in $q_2(z)$, is

$$d = \frac{\frac{3-\sqrt{3}}{4\sqrt{2}}z}{\frac{1-\sqrt{3}}{\sqrt{2}}z} = \frac{3-\sqrt{3}}{4(1-\sqrt{3})} = -\frac{\sqrt{3}}{4} \;.$$

Since $|b_1(z)| = 0$ we know that $r_2(z) = 0$ (the remainder always has degree less than the divisor), so we are looking for $q_2(z)$ such that $a_1(z) = b_1(z)q_2(z)$. Consequently,

$$\left(\frac{1+\sqrt{3}}{4\sqrt{2}} + \frac{3-\sqrt{3}}{4\sqrt{2}}z\right) = \frac{1-\sqrt{3}}{\sqrt{2}}z \cdot \left(cz^{-1} - \frac{\sqrt{3}}{4}\right) \;,$$

which is valid for only one value of c, namely

$$c = \frac{\frac{1+\sqrt{3}}{4\sqrt{2}}}{\frac{1-\sqrt{3}}{\sqrt{2}}} = \frac{1+\sqrt{3}}{4(1-\sqrt{3})} = -\frac{2+\sqrt{3}}{4} \;.$$

Therefore

$$q_2(z) = -\frac{2+\sqrt{3}}{4}z^{-1} - \frac{\sqrt{3}}{4} \;,$$

$$r_2(z) = 0 \;,$$

$$a_2(z) = b_1(z) = \frac{1-\sqrt{3}}{\sqrt{2}}z \;.$$

In order to have the correct high pass filtering, we need to apply (12.27). First we use (12.26) to find H'_{10} and H'_{11}. Note that we use $\frac{1-\sqrt{3}}{\sqrt{2}}z$ as the multiplier in this case.

$$\begin{bmatrix} H_{00}(z) & H_{01}(z) \\ H'_{10}(z) & H'_{11}(z) \end{bmatrix} = \begin{bmatrix} \frac{1-\sqrt{3}}{\sqrt{2}}z & 0 \\ 0 & -\frac{1+\sqrt{3}}{\sqrt{2}}z^{-1} \end{bmatrix} \begin{bmatrix} 1 & \frac{2+\sqrt{3}}{4}z^{-1} + \frac{\sqrt{3}}{4} \\ 0 & 1 \end{bmatrix} \begin{bmatrix} 1 & 0 \\ \sqrt{3} & 1 \end{bmatrix}$$

$$= \begin{bmatrix} \frac{3+\sqrt{3}}{4\sqrt{2}} + \frac{1-\sqrt{3}}{4\sqrt{2}}z & \frac{1+\sqrt{3}}{4\sqrt{2}} + \frac{3-\sqrt{3}}{4\sqrt{2}}z \\ -\frac{\sqrt{3}+3}{\sqrt{2}}z^{-1} & -\frac{1+\sqrt{3}}{\sqrt{2}}z^{-1} \end{bmatrix} \;. \qquad (12.29)$$

We also need $H_{10}(z)$ and H_{11}. Combining (7.35) and (7.66), we find

$$H_1(z) = -cz^{-2k-1}H_0(-z^{-1}) \ . \qquad (12.30)$$

Since in this example

$$H_{00}(z) = h[1] + h[3]z \ , \quad \text{and} \quad H_{01}(z) = h[0] + h[2]z \ , \qquad (12.31)$$

we find by (7.16) that

$$H_0(z) = H_{00}(z^2) + z^{-1}H_{01}(z^2) = h[0]z^{-1} + h[1] + h[2]z + h[3]z^2 \ .$$

With $k = 0$ and $c = 1$ if follows from (12.30) that

$$H_1(z) = -h[0] + h[1]z^{-1} - h[2]z^{-2} + h[3]z^{-3} \ ,$$

and thus

$$H_{10}(z) = -h[0] - h[2]z^{-1} \ , \quad \text{and} \quad H_{11}(z) = h[1] + h[3]z^{-1} \ . \qquad (12.32)$$

We now insert these H_{10} and H_{11} together H'_{10} and H'_{11} from (12.29) into (12.27), which yields (we skip the intermediate calculations)

$$t(z) = H'_{10}(z)H_{11}(z) - H'_{11}(z)H_{10}(z) = (2 + \sqrt{3})z^{-1} \ .$$

Finally, we determine the extra matrix necessary, as shown in (12.28),

$$-(2 + \sqrt{3})z^{-1}\left(\frac{1 - \sqrt{3}}{\sqrt{2}}z\right)^2 = z \ ,$$

(notice again that we use the multiplier $\frac{1-\sqrt{3}}{\sqrt{2}}z$) and then entire $\mathbf{H}(z)$ can now be reconstructed as

$$\mathbf{H}(z) = \begin{bmatrix} \frac{1-\sqrt{3}}{\sqrt{2}}z & 0 \\ 0 & -\frac{1+\sqrt{3}}{\sqrt{2}}z^{-1} \end{bmatrix} \begin{bmatrix} 1 & 0 \\ z & 1 \end{bmatrix} \begin{bmatrix} 1 & -\frac{2+\sqrt{3}}{4}z^{-1} - \frac{\sqrt{3}}{4} \\ 0 & 1 \end{bmatrix} \begin{bmatrix} 1 & 0 \\ \sqrt{3} & 1 \end{bmatrix} .$$

There is still an undesired z in the first matrix, but it can safely be removed. Although the consequence is that the right hand side no longer equals $\mathbf{H}(z)$, it is still a valid factorization into lifting steps. It just results in a different z-transformation of the even and odd low and high pass analysis filters.

12.6 Factorizing Coiflet 12 into Lifting Steps

The next filter has a somewhat longer impulse response, and shows the importance of choosing the right factorization. To avoid filling the following pages with sheer numbers, we always round to four digits or less. This is only in the writing of the numbers, however. The calculation have been performed with several more digits, and more accurate lifting coefficients are given in

Table 12.1. Note that the inverse transform is the right one, up to the number of digits used in the numerical computation, since this is how lifting steps work.

We begin by giving the Coiflet 12 filter taps. They can be found in several software package (but not in *Uvi_Wave*), and in the paper [6, p. 516].

$$\mathbf{h_0} = \begin{bmatrix} 0.0164 & -0.0415 & -0.0674 & 0.3861 & 0.8127 & 0.4170 \\ -0.0765 & -0.0594 & 0.0237 & 0.0056 & -0.0018 & -0.0007 \end{bmatrix}.$$

12.6.1 Constructing the Lifting Steps

We choose a z-transform representation for the odd and even filter taps

$$a_0(z) = \quad 0.0164z^{-2}$$
$$- 0.0674z^{-1} + 0.8127 - 0.0765z^1 + 0.0237z^2 - 0.0018z^3 ,$$
$$b_0(z) = -0.0415z^{-2}$$
$$+ 0.3861z^{-1} + 0.4170 - 0.0594z^1 + 0.0056z^2 - 0.0007z^3 ,$$

and carry out the first step in the algorithm. We choose to match the two z^{-2} terms.

$$q_1(z) = \frac{0.0164}{-0.0415} = -0.3952 ,$$
$$r_1(z) = 0.0852z^{-1} + 0.9775 - 0.1000z^1 + 0.0259z^2 - 0.0021z^3.$$

The next step in the algorithm starts with

$$a_1(z) = -0.0415z^{-2}$$
$$+ 0.3861z^{-1} + 0.4170 - 0.0594z^1 + 0.0056z^2 - 0.0007z^3 ,$$
$$b_1(z) = \quad 0.0852z^{-1} + 0.9775 - 0.1000z^1 + 0.0259z^2 - 0.0021z^3 ,$$

and the quotient $q_2(z)$ is obviously of the form $cz^{-1} + d$. We have three options: Either we match both from the left, both from the right, or c from the left and d from the right. The three cases yield

$$q_2(z) = \frac{-0.0415}{0.0852}z^{-1} + \frac{0.3861 - \frac{-0.0415}{0.0852} \cdot 0.9775}{0.0852}$$
$$= -0.4866z^{-1} + 10.11 , \tag{12.33}$$

$$q_2(z) = \frac{-0.0007}{-0.0021} + \frac{0.3861 - \frac{-0.0007}{-0.0021} \cdot 0.0259}{0.0852}z^{-1}$$
$$= 1.5375z^{-1} + 0.3418 , \tag{12.34}$$

$$q_2(z) = \frac{-0.0415}{0.0852}z^{-1} + \frac{-0.0007}{-0.0021} = -0.4866z^{-1} + 0.3418 , \tag{12.35}$$

respectively. Here we see the problem mentioned at the end of Sect. 12.4.3, namely that some factorizations lead to numerically unstable solutions. In an effort to keep the dynamic range of the coefficients at a minimum, we choose to continue with the numerically smallest $q_2(z)$, i.e. the one in (12.35). In fact, all of the following q's are chosen this way. The next five factorizations are given by

$$
\begin{aligned}
q_2(z) &= -0.4866z^{-1} + 0.3418 \ , \\
r_2(z) &= 0.8326z^{-1} + 0.0342 - 0.0127z - 0.0043z^2 \ , \\
q_3(z) &= 0.1024 + 0.4941z \ , \\
r_3(z) &= 0.5627 - 0.1156z + 0.0325z^2 \ , \\
q_4(z) &= 1.480z^{-1} + 0.3648 \ , \\
r_4(z) &= -0.0187z - 0.016z^2 \ , \\
q_5(z) &= 9.492z^{-1} - 2.017 \ , \\
r_5(z) &= 0.7403 \ , \\
q_6(z) &= -0.0253z - 0.0218z^2 \ , \\
r_6(z) &= 0 \ .
\end{aligned}
$$

Since the next step is setting $b_6(z) = r_6(z) = 0$, we have now reached the first index with b_n equal to zero. Hence, according to Theorem 12.4.2, we have found a greatest common divisor of H_{00} and H_{01}, namely $a_6(z) = b_5(z) = r_5(z) = 0.7403$. This is also the scaling factor, the K in Theorem 12.4.1, as was shown in (12.16) and (12.17).

Inserting now into (12.24)

$$
\begin{aligned}
\begin{bmatrix} H_{00}(z) \\ H_{01}(z) \end{bmatrix} &= \prod_{n=1}^{N/2} \begin{bmatrix} 1 & q_{2n-1}(z) \\ 0 & 1 \end{bmatrix} \begin{bmatrix} 1 & 0 \\ q_{2n}(z) & 1 \end{bmatrix} \begin{bmatrix} K \\ 0 \end{bmatrix} \\
&= \begin{bmatrix} 1 & -.3952 \\ 0 & 1 \end{bmatrix} \begin{bmatrix} 1 & 0 \\ -.4866z^{-1} + .3418 & 1 \end{bmatrix} \begin{bmatrix} 1 & .1024 + .4941z \\ 0 & 1 \end{bmatrix} \\
&\quad \begin{bmatrix} 1 & 0 \\ 1.480z^{-1} + .3648 & 1 \end{bmatrix} \begin{bmatrix} 1 & 9.492z^{-1} - 2.017 \\ 0 & 1 \end{bmatrix} \\
&\quad \begin{bmatrix} 1 & 0 \\ -.0253z - .0218z^2 & 1 \end{bmatrix} \begin{bmatrix} .7403 \\ 0 \end{bmatrix}
\end{aligned}
$$

reproduces the even and odd part of the low pass filter. By substituting

$$
\begin{bmatrix} .7403 \\ 0 \end{bmatrix} \quad \text{with} \quad \begin{bmatrix} .7403 & 0 \\ 0 & (.7403)^{-1} \end{bmatrix} = \begin{bmatrix} .7403 & 0 \\ 0 & 1.351 \end{bmatrix} \ ,
$$

we also get the two filters H'_{10} and H'_{11} in (12.25). These can be converted to the right high pass filter by means of (12.27). In this case we find

$$t(z) = 22.74z^{-1} .$$

We have omitted the intermediate calculations, since they involve the products of large Laurent polynomials.

$$\begin{bmatrix} H_{00}(z) & H_{01}(z) \\ H_{10}(z) & H_{11}(z) \end{bmatrix} = \begin{bmatrix} K & 0 \\ 0 & K^{-1} \end{bmatrix} \begin{bmatrix} 1 & 0 \\ -K^2 t(z) & 1 \end{bmatrix} \prod_{n=M}^{1} \begin{bmatrix} 1 & q_{2n}(z) \\ 0 & 1 \end{bmatrix} \begin{bmatrix} 1 & 0 \\ q_{2n-1}(z) & 1 \end{bmatrix}$$

$$= \begin{bmatrix} .7403 & 0 \\ 0 & 1.351 \end{bmatrix} \begin{bmatrix} 1 & 0 \\ -12.46z^{-1} & 1 \end{bmatrix} \begin{bmatrix} 1 & -.0253z - .0218z^2 \\ 0 & 1 \end{bmatrix}$$

$$\begin{bmatrix} 1 & 0 \\ 9.492z^{-1} - 2.017 & 1 \end{bmatrix} \begin{bmatrix} 1 & 1.480z^{-1} + .3648 \\ 0 & 1 \end{bmatrix}$$

$$\begin{bmatrix} 1 & 0 \\ .1024 + .4941z & 1 \end{bmatrix} \begin{bmatrix} 1 & -.4866z^{-1} + .3418 \\ 0 & 1 \end{bmatrix} \begin{bmatrix} 1 & 0 \\ -.3952 & 1 \end{bmatrix} .$$

Expanding this equation will show that

$$H_{11}(z) = H_{00}(z^{-1}), \quad \text{and} \quad H_{10}(z) = -H_{01}(z^{-1}) ,$$

which was also valid for the Daubechies 4 factorization (compare (12.31) and (12.32)). The equations needed for implementing Coiflet 12 is easily derived from the above matrix equation.

$$d^{(1)}[n] = S[2n+1] - 0.3952\, S[2n] ,$$
$$s^{(1)}[n] = S[2n] - 0.4866\, d^{(1)}[n-1] + 0.3418\, d^{(1)}[n] ,$$
$$d^{(2)}[n] = d^{(1)}[n] + 0.1024\, s^{(1)}[n] + 0.4941\, s^{(1)}[n+1] ,$$
$$s^{(2)}[n] = s^{(1)}[n] + 1.480\, d^{(2)}[n-1] + 0.3648\, d^{(2)}[n] ,$$
$$d^{(3)}[n] = d^{(2)}[n] + 9.492\, s^{(2)}[n-1] - 2.017\, s^{(2)}[n] ,$$
$$s^{(3)}[n] = s^{(2)}[n] - 0.0253\, d^{(3)}[n+1] - 0.0218\, d^{(3)}[n+2] ,$$
$$d^{(4)}[n] = d^{(3)}[n] - 12.46\, s^{(3)}[n-1] ,$$
$$s[n] = 0.7403\, s^{(3)}[n] ,$$
$$d[n] = 1.351\, d^{(4)}[n] .$$

Note that the coefficients in these equations are rounded version of more accurate coefficients, which are given in Table 12.1. The rounded coefficients yield a transformed signal which deviates approximately from 0.1% to 2% from the transformed signal obtained using the more accurate coefficients.

12.6.2 Numerically Unstable Factorization of Coiflet 12

In the previous section we saw the beginning of an unstable factorization of the Coiflet 12 filter. Of the three possible choices of factor $q_2(z)$, we continued

Table 12.1. More accurate coefficients for the Coiflet 12 lifting steps

$d^{(1)}[n]$	-0.3952094886		$d^{(3)}[n]$	9.491856450
$s^{(1)}[n]$	-0.4865531265			-2.017253240
	0.3418203790		$s^{(3)}[n]$	-0.02528002562
$d^{(2)}[n]$	0.1023563847			-0.02182215161
	0.4940618204		$d^{(3)}[n]$	-12.46443692
$s^{(2)}[n]$	1.479728699		$s[n]$	0.7403107249
	0.3648016173		$d[n]$	1.350784159

with the numerically smallest, that is (12.35). To see just how bad a factorization can get, we will now repeat the factorization, this time proceeding with (12.33) instead. Moreover, we choose the left most matching of term each time. The resulting factors are then

$$q_1(z) = -0.3952,$$
$$r_1(z) = \quad 0.08522z^{-1} + 0.9775 - 0.09998z + 0.02590z^2 - 0.002108z^3,$$
$$q_2(z) = \quad 0.4866z^{-1} + 10.11,$$
$$r_2(z) = -9.516 + 0.9641z - 0.2573z^2 + 0.02059z^3,$$
$$q_3(z) = \quad 0.008956z^{-1} - 0.1036,$$
$$r_3(z) = -0.002370z - 0.0005804z^2 + 0.00002627z^3,$$
$$q_4(z) = \quad 4014z^{-1} - 1390,$$
$$r_4(z) = -1.1695z^2 + 0.05710z^3,$$
$$q_5(z) = \quad 0.002027z^{-1} + 0.0005953,$$
$$r_5(z) = -0.000007725z^3,$$
$$q_6(z) = \quad 151381z^{-1} - 7392,$$
$$a_6(z) = -0.000007725z^3.$$

This clearly demonstrates that the factorization algorithm is potentially numerically unstable, and one has to carefully choose which factors to proceed with. Note that although we in the previous section chose the numerically smallest q factor each time, there is a priori no guarantee that this will lead to the most stable solution.

The numerical instability seen here is a well-known aspect of the Euclidean algorithm. The interested reader can look at the examples in the book [8], where some solutions to this problem are also discussed.

Exercises

12.1 Determine the exact number of addition (including all subtractions) and multiplications needed for applying the lifting version of Daubechies 4 to a signal of length $2N$ (disregard any boundary corrections). Compare this to the number of additions and multiplications needed to apply the filter bank version of Daubechies 4 to the same signal.

12.2 Repeat the previous exercise for Coiflet 12.

12.3 Implement the six first lifting steps (those originating from the q's) for both the numerically stable and unstable Coiflet 12 in MATLAB (or some other language), and apply it to a random signal. Plot the intermediate signals in the lifting steps to examine how good or bad each of the two factorizations are.

12.4 The CDF(5,3) equation are given by (see [27])

$$d^{(1)}[n] = S[2n+1] - \frac{1}{5}S[2n] ,$$
$$s^{(1)}[n] = S[2n] - \frac{1}{24}(15d^{(1)}[n-1] + 5d^{(1)}[n]) .$$
$$d^{(2)}[n] = d^{(1)}[n] - \frac{1}{10}(15s^{(1)}[n] + 9s^{(1)}[n+1]) ,$$
$$s^{(2)}[n] = s^{(1)}[n] - \frac{1}{72}(-5d^{(2)}[n-1] - 24d^{(2)}[n] + 5d^{(2)}[n+1]) ,$$
$$\tilde{s}[n] = 3\sqrt{2}s^{(2)}[n] ,$$
$$\tilde{d}[n] = \frac{\sqrt{2}}{6}d^{(2)}[n] .$$

Construct the corresponding lifting steps in 2×2 matrix form.

12.5 The Symlet 10 is an orthogonal filter, and the IR is given by

1	0.01953888273525	6	0.72340769040404
2	-0.02110183402469	7	0.19939753397685
3	-0.17532808990805	8	-0.03913424930231
4	0.01660210576451	9	0.02951949092571
5	0.63397896345679	10	0.02733306834500

The coefficients can also be generated using **symlets** from *Uvi_Wave* in MATLAB. Convert this filter into lifting steps in the equation form.

13. Wavelets in Matlab

In the previous chapters we have presented wavelets through the lifting technique. We have also shown that this approach is equivalent with the approach based on filters. This means that one can use any of the available software packages (all based on the filter approach) to experiment with wavelet analysis.

In this chapter we present a number of examples using MATLAB and the *Uvi_Wave* toolbox. We assume limited familiarity with MATLAB. All the examples here can be type directly at the MATLAB prompt, and the results can be studied without any knowledge of MATLAB. In some of the previous chapters we gave a number of exercises requiring MATLAB, and the techniques learned in this chapter are needed to solve these. Although a number of useful MATLAB commands are presented in the examples, not everything can learned from this chapter, and we refer to any book on basic MATLAB programming for further study. We also assume some familiarity with concepts from signal processing. As a consequence, we use the term 'number of filter taps' for the length of a filter.

The *Uvi_Wave* toolbox is one of many wavelet software package. We have chosen this particular one mainly because it is of limited scope, but contains most of the functions we need. The toolbox is in the public domain and can be acquired electronically, see Chap. 14. Our examples can easily be adapted to the WaveLab toolbox and to the official MATLAB wavelet toolbox by anyone familiar with these toolboxes. For information on these two toolboxes, see the references in Chap. 14.

We recommend the reader to go through the examples, typing in the steps at the MATLAB prompt. Alternatively, the examples are available electronically, see Chap. 14. We also encourage the reader to do further experiments, using the methods introduced in the examples.

We have shown the graphical output of many, but not all, of the examples, to guide the reader through the examples. Some of the figures have been reduced in size to fit on the page.

We have presented the graphical output from *Uvi_Wave* unchanged. Unfortunately, the visualization tools in this toolbox do not present the results in a unified manner. The wavelet transform of a signal is plotted with the first level at the top, and the last level at the bottom. In the output from the multiresolution function this order is reversed. For the wavelet packets the situation is even more complicated. The result of a full wavelet decomposition is a matrix, where each level is stored in a column. The first column is the original signal, and the last column the final level permitted by a given signal. However, in graphical output the original signal is at the top, and the last level at the bottom. So it is important that the reader consults the documentation for the various function to find out how the output is stored in a vector or a matrix.

Due to changes in MATLAB in versions 5.x, some functions in version 3.0 of *Uvi_Wave* produce errors and warnings. We suggest how to fix these problems, see Chap. 14.

13.1 Multiresolution Analysis

In the following examples we use line numbers of the form 1.1, where the first number refers to the example being considered, and the second number gives the line numbers within this example.

We start with a signal consisting of a sine sampled 32 times per cycle. The signal is 500 samples long.

```
1.1 >   S = sin([1:500]*2*pi/32);
```

It is easy to show that there are 32 samples per cycle.

```
1.2 >   plot(S(1:32))
1.3 >   plot(S)
```

To carry out a wavelet analysis of this signal, we must choose four filters. In Chap. 7 the notation for the analysis filter pair was h_0, h_1, and for the synthesis pair g_0, g_1. We changed this notation to h, g for the analysis pair, and $\tilde{h}$, $\tilde{g}$ for the synthesis pair in Chap. 8. Since we cannot easily use the $\tilde{h}$ notation in MATLAB, we change the notation once more, this time to h, g, rh, and rg. This change in notation also corresponds to the notation used in the *Uvi_Wave* documentation. Filters are generated by several different functions in *Uvi_Wave*. The members of the Daubechies family of orthogonal filters are generated by the function daub. It needs one argument, which is the length of the filter, or the number of filter taps. We start our experiments with the Daubechies 8 filters, which are generated by the command

```
1.4 >   [h,g,rh,rg] = daub(8);
```

The easiest way to see the result of a multiresolution analysis (MRA) of the signal S is to use the *Uvi_Wave* function multires, which produces a series of graphs similar to those shown in Fig. 4.4 (see Sect. 4.1 for an explanation of the concept of an MRA). With the help function it is possible to get a description of what multires does, and what parameters are needed. It takes the signal, the four filters, and the finally the number of levels we want in the decomposition. Here we choose to use 4 levels.

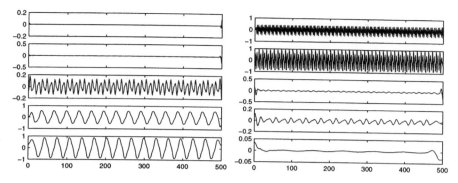

Fig. 13.1. Graphical output from line 1.9 (left) and line 1.13 (right)

```
1.5 >  help multires
1.6 >  y = multires(S,h,rh,g,rg,4);
1.7 >  size(y)
```

By typing `size(y)`, the size of the matrix `y` is returned. It has 5 rows and 500 columns. With the function `split` each of the 5 rows is shown in the same figure, but with vertical axes each having its own scale. The horizontal axes are all identical, running from 0 to 500, the sample indices. The result of using `split` on `y` is shown on the left in Fig. 13.1. To simplify the figure we have removed some redundant labels. Note how most of the energy is concentrated in the two bottom graphs. We recall that the function `split` plots the first level at the bottom and the last level at the top of a figure.

```
1.8 >  help split
1.9 >  split(y)          (see Fig. 13.1)
```

A sine with a higher frequency (5 samples per cycle) has a different energy distribution in the 5 rows of the decomposition.

```
1.10 >  S = sin([1:500]*2*pi/5);
1.11 >  y = multires(S,h,rh,g,rg,4);
1.12 >  figure
1.13 >  split(y)          (see Fig. 13.1)
```

While the first (low frequency) signal with 32 samples per cycle has the main part of its energy in the two bottom rows, the second (high frequency) signal has most of its energy in the two top rows, see the right part of Fig. 13.1. This shows one of the features of a decomposition, namely the ability to split a signal according to frequency.

Instead of using just a sine, we can try to add a couple of transients, i.e. a few samples deviating significantly from their neighbor samples.

```
2.1 >  close all
2.2 >  S = sin([1:512]/512*2*pi*5);
```

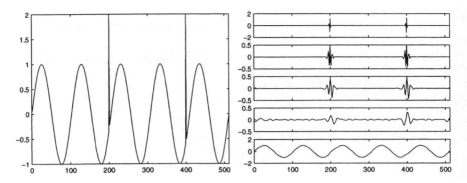

Fig. 13.2. Graphical output from line 2.5 (left) and line 2.7 (right)

```
2.3 >   S(200) = 2;
2.4 >   S(400) = 2;
2.5 >   plot(S)          (see Fig. 13.2)
```

This time the signal is a sine with 5 cycles sampled 512 times. Samples number 200 and 400 are set to 2. We again perform an MRA on the signal.

```
2.6 >   y = multires(S,h,rh,g,rg,4);
2.7 >   split(y)          (see Fig. 13.2)
```

The signal S and the MRA of the signal are shown in Fig. 13.2. The bottom graph in the MRA is hardly distinguishable from the original sine, while the four other graphs contain most of the energy from the transients. This property of the MRA allows us to separate S into the transients, and the sine. First we note that since the wavelet transform is linear, it is possible to reconstruct the original signal S simply by adding all the rows in the MRA. This fact can also be found in the help text to multires. It is easy to check this graphically.

```
2.8 >   figure
2.9 >   plot(sum(y,1)-S)
```

Typing sum(y,1) adds all entries along the first dimension in y, which is equivalent to adding all the rows in y. It can also be done by the more intuitive, but cumbersome, command

```
2.10 >   plot(y(1,:)+y(2,:)+y(3,:)+y(4,:)+y(5,:)-S)
```

It is clear that the original signal and the reconstructed signal are almost identical. They are not completely identical due to the finite precision of the computations on the computer.

Since the bottom graph resembles the original sine, it should be possible to reconstruct the transients from the remaining four parts of the decomposition. The following commands

```
2.11 >  figure
2.12 >  plot(y(1,:)+y(2,:)+y(3,:)+y(4,:))
```

show that the transients are fairly well separated from the sine in the decomposition. We can also plot the bottom graph in the same figure for comparison.

```
2.13 >  hold on
2.14 >  plot(y(5,:))
```

Until now we have only been looking signals reconstructed from wavelet coefficients (since the signals in the plots generated with multires followed by split are not the wavelet coefficients themselves, but reconstructions of different parts of the coefficients). If we want to look at the wavelet coefficients directly, the function wt can be used. It implements exactly what is depicted in both Fig. 3.7 and Fig. 8.2(a). Therefore by typing

```
2.15 >  yt = wt(S,h,g,4);
```

the signal S is subjected to a 4 scale DWT (based on the Daubechies 8 filters). Although the output of wt is a single vector, it actually contains 5 vectors. How these are located in yt are described in the help text to wt. It is also shown in Fig. 4.2. The wavelet coefficients can easily be shown with the isplit function. Note that this function plots the first level at the top, in contrast to split, which plots the output of multires in the opposite order.

```
2.16 >  figure
2.17 >  isplit(yt,4,'','r.')
```

Because we now have the wavelet coefficients available, we can experiment with changes to the coefficients, for example setting some of them equal to zero. After changing the coefficients, the function iwt is used to do an inverse transform. Suppose we want to see what happens, if the fourth scale coefficients are set to zero (the fourth scale coefficient vector is denoted by d_{j-4} in Fig. 3.7). Then we use the commands

```
2.18 >  yt(33:64) = zeros(1,32);
2.19 >  yr = iwt(yt,rh,rg,4);
```

With subplot two or more graphs can be inserted into the same figure (the same window), making it easier to compare them. Here the first graph shows the original signal in blue and the reconstructed signal (from the modified coefficients) in red.

```
2.20 >  figure
2.21 >  subplot(211)
2.22 >  plot(S,'b')
2.23 >  hold on
2.24 >  plot(yr,'r')
```

As the second graph the difference between the two signals in the first subplot is shown.

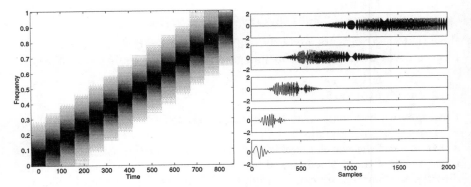

Fig. 13.3. Graphical output from line 3.3 (left) and line 3.7 (right)

```
2.25 >   subplot(212)
2.26 >   plot(S-yr)
```

13.2 Frequency Properties of the Wavelet Transform

Before going through this section the reader should be familiar with the concept of a time-frequency plane from Chap. 9, and the function specgram from the signal processing toolbox for MATLAB.

We have a number of times stressed that **h** and **g** are low and high pass filters, respectively, i.e. they can separate (more or less) the low and high frequencies in a signal. To see what effect this property has on the wavelet transform, we will now use multires on a signal containing all frequencies from 0 to half the sampling frequency.

```
3.1 >   close all
3.2 >   S = sin([1:2000].^2/1300);
```

With a Fourier spectrogram we immediately see that the signal, a so-called chirp, actually does contain all frequencies.

```
3.3 >   specgram(S)    (see Fig. 13.3)
```

The filter is chosen to be a 16 tap filter, i.e. a filter of length 16, from the Daubechies family, in order to have a reasonable frequency localization.

```
3.4 >   [h,g,rh,rg]=daub(16);
3.5 >   y = multires(S,h,rh,g,rg,4);
3.6 >   figure
3.7 >   split(y)       (see Fig. 13.3)
```

The left part of Fig. 13.3 shows that the energy is distributed along a line in the time-frequency plane. This agrees with the linear dependence between time and frequency in a chirp, which here is obtained by sampling the function $\sin(t^2/1300)$. The MRA graphs on the right in Fig. 13.3 do not show a linear dependence. Approximately half of the energy is located in the top graph, a quarter in the second graph, and so on. Each graph seems to contain about half of the energy of the one above. This partitioning of the energy comes from the repeated use of the filters on the low pass part in each step of the DWT. As a consequence, the relation between time and frequency becomes logarithmic, not linear.

As another example of the ability of the DWT to separate frequencies, we look at a signal mixed from four signals, each containing only one frequency, and at different times. To experiment with another kind of filter, we choose to use a biorthogonal filter, the CDF(3,15) filter, which is obtained from

```
4.1 >   [h,g,rh,rg]=wspline(3,15);
```

As explained in Chap. 7, biorthogonal filters do not necessarily have an equal number of low and high pass filter taps.

```
4.2 >   h
4.3 >   g
```

For our experiment we want four different signals (different frequency content at different times)

```
4.4 >   s1 = [sin([1:1000]/1000*2*pi*20) zeros(1,1000)];
4.5 >   s2 = [zeros(1,1000) sin([1:1000]/1000*2*pi*90)];
4.6 >   s3 = [zeros(1,1200) sin([1:400]/400*2*pi*2) ...
            zeros(1,400)];
4.7 >   s4 = [zeros(1,250) ...
            sin([1:250]/250*2*pi*125+pi/2) zeros(1,250) ...
            sin([1:500]/500*2*pi*250+pi/2) zeros(1,750)];
```

where s4 contains a high frequency, s3 contains a low frequency, while s1 and s2 have frequencies in between. This can be verified by plotting the four signals. The plot also shows how the energy is distributed in time.

```
4.8 >    subplot(511)
4.9 >    plot(s1)
4.10 >   subplot(512)
4.11 >   plot(s2)
4.12 >   subplot(513)
4.13 >   plot(s3)
4.14 >   subplot(514)
4.15 >   plot(s4)   (see Fig. 13.4)
```

We combine the four signals by addition.

```
4.16 >   S = s1+s2+s3+s4;
```

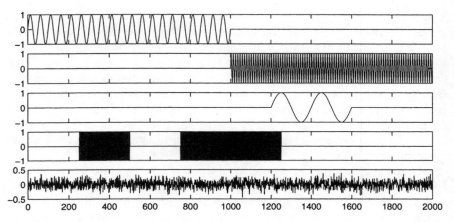

Fig. 13.4. Graphical output from lines 4.8 to 4.15, and from line 4.19

To see how the DWT reacts to noise, we will test the MRA on this signal
S both with, and without, noise. Normal distributed random numbers are a
common type of simulated noise.

```
4.17 >   s5 = randn(1,2000)/8;
```

The probability that **randn** generates numbers with absolute value larger
than 4 is very small. Division by four leads to a signal **s5** with almost all its
values between −0.5 and 0.5.

```
4.18 >   subplot(515)
4.19 >   plot(s5)   (see Fig. 13.4)
```

Thus we have the following two signals for our experiments.

```
4.20 >   figure
4.21 >   subplot(211)
4.22 >   plot(S)
4.23 >   subplot(212)
4.24 >   plot(S+s5)
```

Let us first look at the MRA for the signal S without noise.

```
4.25 >   ym1 = multires(S,h,rh,g,rg,5);
4.26 >   figure
4.27 >   split(ym1)
```

When interpreting the graphs, it is essential to notice that they have different
vertical scales. If we want the same scale on all graphs, the function **set(gca,
...)** is useful (it can be used to alter graphs in many other ways, too).

```
4.28 >   for n=1:6     (see Fig. 13.5)
             subplot(6,1,n)
```

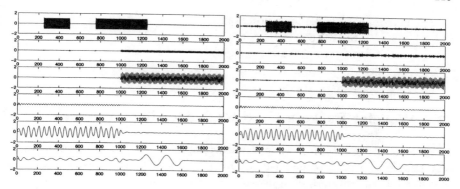

Fig. 13.5. Graphical output from line 4.28 and line 4.40

```
      set(gca, 'YLim', [-2 2])
   end
```

Now all the graphs are scaled to the interval $[-2; 2]$.

Since the 'up'-key on the keyboard can be used to browse through previous lines, it would be nice to have the for loop on one line (then the for loop can easily be reused).

```
4.29 >  for n=1:6, subplot(6,1,n), set(gca, 'YLim', [-2 2]), end
```

For clarity, subsequent loops will not be written in one line, though.

Now we can see one of the major advantages of the DWT: Not only have the four frequencies been separated, but the division in time is also reconstructed (compare Fig. 13.4 and Fig. 13.5). If we try to separate the frequencies using the short time Fourier transform (which forms the basis for the Fourier spectrogram), we have to choose between good frequency localization

```
4.30 >  figure
4.31 >  subplot(211)
4.32 >  specgram(S, 2048, 1, 256)
4.33 >  caxis([-50 10])
```

and good time localization.

```
4.34 >  subplot(212)
4.35 >  specgram(S, 2048, 1, 32)
4.36 >  caxis([-50 10])
```

The MRA for S+s5, i.e. S with noise added, is roughly the same as the MRA for S.

```
4.37 >  ym2 = multires(S+s5,h,rh,g,rg,5);
4.38 >  figure
4.39 >  split(ym2)
```

```
4.40 >   for n=1:6    (see Fig. 13.5)
             subplot(6,1,n)
             set(gca, 'YLim', [-2 2])
         end
```

13.3 Wavelet Packets Used for Denoising

As an example of a typical wavelet packets application, we will now study a very simple approach to denoising. This example demonstrates advantages, as well as disadvantages, of applying a wavelet based algorithm. Usually the noise-free signal is not available a priori, but to evaluate the effectiveness of our technique, we use a synthetic signal, to which we add a known noise.

```
5.1 >   S = sin([1:4096].^2/5200);
5.2 >   specgram(S,1024,8192,256,192)
```

First a remark on the use of colors. By typing

```
5.3 >   colorbar
```

another axis appears to the right in the figure. This axis shows how the colors are distributed with respect to the numerical values in the spectrogram. Sometimes the color scale interval is inappropriate, and it can be necessary to change it. In the present case the interval seems to be $[-150; 30]$, and it can be changed using caxis in the following way.

```
5.4 >   caxis([-35 35])
5.5 >   colorbar
```

The caxis command changes only the colors in the main plot. So if the colorbar should correspond to the spectrogram, it has to be updated by reissuing the command colorbar.

The chirp is chosen as the synthetic signal here, since it covers a whole range of frequencies, and any noise in such a signal is easily seen and heard. The latter being possible only if the computer is equipped with a sound card. The signals can be played using sound, which takes the playback frequency as second argument.

```
5.6 >   sound(S,8192)
```

Being a synthetic signal, S does not have a 'real' sampling frequency. In this case we just chose 8 kHz. The input signal to sound must have value in $[-1; 1]$. Values outside this interval are clipped, so any signal passed to sound should be scaled to fit this interval. Note that our synthetic signal was created using the sine, hence no scaling is necessary.

As noise we choose 4096 randomly distributed numbers.

```
5.7 >   noise = randn(1,4096)/10;
5.8 >   figure
5.9 >   specgram(S+noise,1024,8192,256,192)
5.10 >  caxis([-35 35])
5.11 >  figure
5.12 >  specgram(noise,1024,8192,256,192)
5.13 >  caxis([-35 35])
5.14 >  sound(S+noise,8192)
```

The intensity of the noise can be varied by choosing a number different from 10 in the division in line 5.7. As our filter we can choose for example a member of the CDF family (often called spline wavelets)

```
5.15 >  [h,g,rh,rg]=wspline(3,9);
```

or near-symmetric Daubechies wavelets, from the symlet family.

```
5.16 >  [h,g,rh,rg]=symlets(30);
```

Our experiments here use the symlets. Since symlet filters are orthogonal, the transform used here preserves energy.

To make a wavelet packet decomposition, the function wpk is used. Since it operates by repeatedly applying wt, and since it always does a full decomposition, it is relatively slow. This does not matter in our experiments here. But if one needs to be concerned with speed, then one can use decomposition functions implemented in C or FORTRAN, or, as we will see later, one can specify which basis to find. Some results on implementations are given in Chap. 11, and information on available libraries of functions for wavelet analysis is given in Chap. 14. The function wpk is sufficient for our experiments.

```
5.17 >  y = wpk(S,h,g,0);
```

The fourth argument determines the ordering of the elements on each level. The options are filter bank ordering (0) and natural frequency ordering (1). The meaning of these terms is described in Sect. 9.3. In this case the ordering does not matter, since we are only concerned with the amplitude of the wavelet coefficients (in line 5.29 small coefficients are set equal to zero). With wpk we get a decomposition of $\lfloor \log_2(N) \rfloor + 1$ levels, where N is the length of the signal. So wpk(S,h,g,0) returns a 4096×13 matrix, since $\lfloor \log_2(4096) \rfloor + 1 = 13$, and the first column of this matrix contains the original signal. Note that the ordering of the graphs in a plot like Fig. 13.6 corresponds to the transpose of the y obtained from wpk. The ordering of elements in y is precisely the one depicted in Fig. 8.2(b).

At the bottom level (i.e. in the last column of y) the elements are quite small, only one coefficient wide. Since they are obtained by filtering the elements in the level above (where each element is two coefficients wide) only

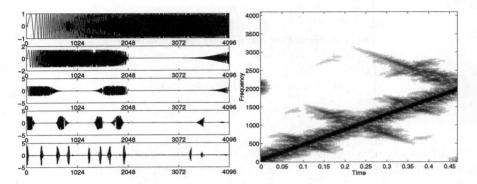

Fig. 13.6. Graphical output from line 5.19 (with symlets 30 from line 5.16) and line 5.32

2 of the 30 filter taps are used. Hence 28 of the filters taps do not have any influence on the bottom level, and one should therefore be careful when interpreting the lowest level. This also applies (in lesser degree) to the few levels above it. However, it is still possible to reconstruct the original signal from these elements, so the lowest levels does give a representation of the original signal, although it might not be useful in some applications.

A plot of all 13 levels in the same figure would be rather cluttered. We limit the plot to the first 5 levels here. Note that wpk produces a decomposition in a matrix, where each column corresponds to a level. With set(gca,...) the time axis is set to go from 0 to 4096, and tick marks are applied at 0, 1024, 2048, 3072, and 4096. On the fifth level there are 16 elements (see Fig. 13.6), four between each pair of tick marks.

```
5.18 >   figure
5.19 >   for n=1:5        (see Fig. 13.6)
             subplot(5,1,n)
             plot(y(:,n))
             set(gca, 'XLim', [0 4096], 'XTick', [0:1024:4096])
         end
```

Visually, the decomposition of the noisy signal does not deviate much from the first decomposition.

```
5.20 >   yn = wpk(S+noise,h,g,0);
5.21 >   figure
5.22 >   for n=1:5
             subplot(5,1,n)
             plot(yn(:,n))
             set(gca, 'XLim', [0 4096], 'XTick', [0:1024:4096])
         end
```

There is nonetheless an important difference (on all levels but for the sake of simplicity we focus on the fifth level): While most of the fifth level in the decomposition of the signal without noise consists of intervals with coefficients very close to zero, the fifth level in the decomposition of the noisy signal is filled with small coefficients, which originate from the noise we added to the signal.

This gives rise to an important observation. The energy of the signal is collected in fewer coefficients as we go down in the levels. Energy is conserved, since we have chosen orthogonal filters. Consequently, these coefficients must become larger. The noise, however, since it is random, must stay evenly distributed over all levels. Thus due to the energy preservation, most of the coefficients coming from the noise must be small. The growth of the coefficients is clearly visible in Fig. 13.6 (note the changed scaling on the vertical axes). It is therefore reasonable to hope that the signal can be denoised by setting the small coefficients equal to zero. The property of concentrating desired signal information without concentrating the noise is important in many applications.

We have already looked at level 5, so let us look at level 7, for example. We plot level 7 from the two decompositions in the same figure.

```
5.23 >  figure
5.24 >  plot(y(:,7))
5.25 >  hold on
5.26 >  plot(yn(:,7),'r')
```

Since there are 4096 points in each signal, the graphs do not clearly show the differences. By typing

```
5.27 >  zoom
```

(or by choosing zoom on the figure window menu) we can enlarge a chosen area of the graph. Mark an area with the mouse, and MATLAB zooms to this area. By examining different parts of the signals, one gets the impression that the difference between the two signals is just the noise. Hence after six transforms the noise remains as noise. This examination also reveals that setting all coefficients below a certain threshold (between 0.5 and 1) equal to zero will make the two signal much more alike. To try to determine the best threshold, we look at the coefficients on the seventh level order according to absolute value.

```
5.28 >  figure; plot(sort(abs(yn(:,7))))
```

The choice of threshold value is not obvious. We have chosen 1, but judging from the sorted coefficients, this might be a bit to high. To change all values in yn(:,7) with absolute value less than or equal to 1 to zero we type

```
5.29 >  yc = yn(:,7) .* (abs(yn(:,7)) > 1);
```

The part abs(yn(:,7))>1 returns a vector containing 0's and 1's. A 0, whenever the corresponding value is below or equal to 1, and 1 otherwise. Multiplying coefficient by coefficient with yn(:,7) leaves all coefficients above 1 unchanged, while the rest is changed to zero. Note that .* is coefficient-wise multiplication and * is matrix multiplication.

Now we want to used the modified seventh level to reconstruct a hopefully denoised signal. This is done using iwpk. Since this is an inverse transform, we need to use the synthesis filters (rh and rg).

```
5.30 >   yr = iwpk(yc,rh,rg,0,6*ones(1,64));
```

The fifth argument passed to iwpk is the basis to be used in the reconstruction. As we saw in Chap. 8, many possibilities exist for the choice of a basis (or a representation), when we use a full wavelet packet decomposition. Here we have chosen the representation given by all elements on the seventh level. This basis is in *Uvi_Wave* represented as a vector of length 64 consisting of only 6's (given as 6*ones(1,64) in MATLAB), where 6 is the level (counting from zero) and 64 the number of elements on the seventh level. The basis representation notation is described by Fig. 11.2 on page 182, and in *Uvi_Wave* by typing basis at the prompt. The reconstruction in line 5.30 is performed exactly as shown in Fig. 8.3 on page 90.

Now yr contains a reconstructed, and hopefully denoised, signal. Let us look at the spectrogram of this signal.

```
5.31 >   figure
5.32 >   specgram(yr,1024,8192,256,192) (see Fig. 13.6)
5.33 >   caxis([-35 35])
5.34 >   sound(yr, 8192)
```

We can visualize our denoising success by looking at the difference between the original signal and the denoised signal. Ideally, we should only see noise. Often a (small) part of the signal is also visible in the difference. This means that we also have removed a (small) part of the signal.

```
5.35 >   figure
5.36 >   specgram((S+noise)-yr',1024,8192,256,192)
5.37 >   caxis([-35 35])
5.38 >   sound(yr'-(S+noise), 8192)
```

Finally, we can inspect the difference between the original, clean, signal, and the denoised signal. This is only possible, because the signal in our case is synthetic; the point of denoising a signal is usually that the clean signal is not available. Having the clean signal available gives us a chance to examine the efficiency of our denoising algorithm.

```
5.39 >   figure
5.40 >   specgram(yr'-S,1024,8192,256,192)
5.41 >   caxis([-35 35])
5.42 >   sound(yr'-S), 8192)
```

To experiment with a different threshold value, some of the calculations must be performed again. By collecting these in a loop, it is possible to try several threshold values without having to retype the commands. Note that this loop only works if **yn** is already calculated.

```
5.43 >   close all
5.44 >   while 1
           Bound = input('Bound (return to quit): ');
           if isempty(Bound) break; end
           yc = yn(:,7) .* (abs(yn(:,7)) > Bound);
           yr = iwpk(yc,rh,rg,0,6*ones(1,64));
           figure(1)
           clf
           specgram(yr,1024,8192,256,192)
           caxis([-35 35])
           sound(yr,8192)
         end
```

13.4 Best Basis Algorithm

Wavelet packet decompositions are often used together with the best basis algorithm to search for a best basis, relative to a given cost function. These topics were discussed in detail in Chap. 8.

In *Uvi_Wave* there are several functions that search through the full wavelet packet decomposition. Only one of them, **pruneadd**, implements the best basis search algorithm, as described in Chap. 8 on page 94. We will use this function for our next example.

We start by defining a signal. A rather short one is chosen this time, since we are interested in basis representations, and since the number of representations grows rapidly with the length of the signal, when we perform a full wavelet packet decomposition.

```
6.1 >   S = sin([1:32].^2/30);
6.2 >   S = S / norm(S);
6.3 >   plot(S)
```

The signal is normalized to have norm equal to one, because we want to use Shannon's entropy cost function. Although it is not necessary to do this (the resulting basis would be the same without normalization), it ensures positive cost values.

```
6.4 >   [h,g,rh,rg] = daub(6);
6.5 >   [basis,v,total] = pruneadd(S,h,g,'shanent');
```

The function **pruneadd** does several things. As input is takes the signal, the two analysis filters, and a cost function. It first performs a full wavelet packet decomposition of the given signal, to the number of levels permitted by the length of the signal. In our case we have a signal of length 32, so there will be 6 levels in the decomposition (see Table 8.2). The output is the selected basis, contained in

6.6 > basis

the representation of the signal in this basis, i.e. the coefficients in v,

6.7 > plot(v)

and the total cost for that particular representation.

6.8 > total

The cost function **shanent** calculates Shannon's entropy for the output vector v (see Sect. 8.3.2).

6.9 > shanent(v)

This value is the one returned by **pruneadd** as **total** above.

Since there is no function in *Uvi_Wave*, which performs only the best basis search, we will now show that this search algorithm is very easy to implement in MATLAB. Note how each step in the algorithm on page 94 can be converted to MATLAB code. First we need a full decomposition of the signal.

6.10 > y = wpk(S,h,g,0);
6.11 > size(y)

The first step in the algorithm is to calculate the cost value for each element in the decomposition. Since there are 6 levels, each of length 32, there is a total of $2^6 - 1 = 63$ elements. We start by constructing a vector containing the computed cost values.

```
6.12 >  CostValue = [ ];
6.13 >  for j=0:5
           for k=0:2^j-1
             Element = y(1+k*2^(5-j):(k+1)*2^(5-j),j+1);
             CostValue = [CostValue shanent(Element)];
           end
         end
```

Note that this construction of the vector CostValue is highly inefficient, and it is used here only for the sake of simplicity. Now CostValue contains the 63 cost values from our decomposition. Save the cost values for later use (if one omits the ; at the end, the values are shown on the display).

6.14 > Old_CostValue = CostValue

Since we know the cost value for the best basis for our example above, we can easily check, if the best representation happens to be the original signal.

```
6.15 >  CostValue(1)-total
```

This is not the case, since the original representation has a higher cost than the value found by **pruneadd** above. The next step in the algorithm is to mark all the elements at the bottom level. We now need a notation for a basis. The one used in *Uvi_Wave* is efficient, but not user friendly. We continue to use the notation implicitly given by the definition of the vector CostValue. Here we have numbered elements consecutively, from the top to the bottom in each column, and then going through the rows from left to right. The last column contains level 6, which has 32 elements. This numbering is performed in the two **for** loops starting on the line 6.13. It is also depicted to the right in Fig. 11.2 on page 182. In the basis vector we let 1 mean a chosen (marked) element and 0 marks elements not chosen. Marking all the elements at the bottom level is then performed by

```
6.16 >  b = [zeros(1,31) ones(1,32)];
```

With 6 levels, there is a total of 63 elements, distributed with 31 elements at the first 5 levels and 32 elements at the bottom level. In the next steps the actual bottom-up search is carried out. Note that for any element with index equal to Index the two elements just below it have indices 2*Index and 2*Index+1.

```
6.17 >  Index = 31;
6.18 >  for j = 4:-1:0
           for k = 0:2^j-1
              tmp = CostValue(2*Index)+CostValue(2*Index+1);
              if CostValue(Index) < tmp
                 b(Index) = 1;
              else
                 CostValue(Index) = tmp;
              end
              Index = Index - 1;
           end
        end
```

Note that here we do not remove the marks on elements below the currently chosen and marked ones, in contrast to the step 4(a) in the algorithm. We will do this step later. Before proceeding with the basis, let us take a look at the cost values. As can be seen in the algorithm, step 4(b), the numbers in CostValue might change.

```
6.19 >  Old_CostValue - CostValue
```

Most of the numbers in the vector CostValue have changed as part of the process of finding the best basis. The last step in the algorithm (on page 94) states that the first entry in CostValue is the total cost value for the basis just found, which is the best basis.

```
6.20 >  CostValue(1)-total
```

If everything has gone as expected, you should get zero. Only one thing remains to be done. Since one step was neglected in the best basis search above, there are 'too many' 1's in b, because the 1's that should have been removed by step 4(a), are still in the vector. These can be removed by

```
6.21 >  for j=1:31
            b(2*j) = b(2*j) + 2*b(j);
            b(2*j+1) = b(2*j+1) + 2*b(j);
        end
6.22 >  b = (b == 1);
```

The for loop sets all the unwanted 1's to values higher than 1, and the last line sets all values but 1 to zero.

The two vectors basis and b now represent the same basis, but in two different ways. While basis is a 'left to right' representation (see Sect. 11.8 or the script basis in *Uvi_Wave*) the vector b is a 'top-down' representation. Both basis representations are shown in Fig. 13.7.

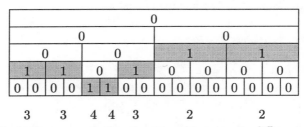

Fig. 13.7. The basis representations basis and b are quite different, and yet they represent the same basis

It is not difficult to reconstruct the original signal from v, which is the representation of the signal in the basis basis. The process is exactly the one described by Fig. 8.3.

```
6.23 >  S2 = iwpk(v,rh,rg,0,basis);
6.24 >  plot(S2-S)
```

The best basis can be presented using the two commands tree and tfplot. They both take a basis as input arguments. Both types of representations have been demonstrated in previous chapters. The former relates to the type of visualization presented in Sect. 3.5, although only wavelet decomposition is discussed there. The tfplot command shows the corresponding time-frequency plane without any coloring, like the time-frequency planes in Fig. 9.11.

```
6.25 >  tree(basis)
```

```
6.26 >   figure
6.27 >   tfplot(basis)
```

Place the two figures next to each other on the screen, and notice how the two representations are much alike, although they might seem different.

The representation of a signal in the best basis (in this case v) is in many application the very reason for using wavelet analysis. Subject to the given cost function, it is the best representation of the signal, and with the right choice of cost functions, we have a potentially very good representation. For instance one with only very few large sample.

In this case we have the representation with the smallest Shannon entropy (since we used the argument 'shanent' in line 6.5). This cost functions finds the representation with the lowest entropy. Since entropy, as explained in Sect. 8.3.2, measures concentration of energy, we have found the representation v with the highest concentration. This in turn means many coefficients in v must be small.

```
6.28 >   figure
6.29 >   plot(sort(abs(v)))
```

Compare this to the size of the coefficients of the original signal.

```
6.30 >   hold on
6.31 >   plot(sort(abs(S)),'r')
```

We have previously experimented with altering the transformed signal in an attempt to denoise a signal (for instance in line 5.29), but so far we have only considered altering the signal on a single level. With a best basis representation we are more in control of what happens to the signal, since we alter a predetermined best representation instead of a more or less arbitrary level representation. This is not so easy to see with just 32 coefficients, so we take a longer signal. We also make some noise.

```
6.32 >   S = sin([1:512].^2/512);
6.33 >   noise = randn(1,512)/2;
```

In all the previous examples we have used randomly distributed noise only. This time we will use colored noise. It is made by band pass filtering the noise, and we use the following command to make this band pass filter. Note that butter comes from the signal processing toolbox.

```
6.34 >   [B,A] = butter(2,[0.2 0.6]);
```

It lets frequencies from 0.2 to 0.6 times half the sampling frequency through. The filter is applied to the noise and added to the signal as follows,

```
6.35 >   S2 = S + filter(B,A,noise);
```

Now we make both a wavelet packet decomposition and a best basis search.

```
6.36 >  [h,g,rh,rg] = daub(10);
6.37 >  y = wpk(S2,h,g,0);
6.38 >  [basis,v]=pruneadd(S2,h,g,'shanent');
```

We see that the best basis is not just a single level.

```
6.39 >  tree(basis)
```

We will now display the sorted coefficients of the original signal, of all the levels in the WP decomposition, and of the best basis representation of the signal.

```
6.40 >  plot(sort(abs(S2)),'b')
6.41 >  hold on
6.42 >  plot(sort(abs(y(:,2:end))),'k')
6.43 >  plot(sort(abs(v)),'r')
```

As stated above the best basis representation is more concentrated than any single level (although in this particular case not so much).

13.5 Some Commands in Uvi_Wave

To encourage the reader to do further experiments with MATLAB and *Uvi_Wave*, we wrap up this chapter with a list of descriptions of useful commands in *Uvi_Wave*, including those presented previously in this chapter. Whenever the theoretical background of a command can be found in the book, we also provide a reference to that particular chapter or section.

This collection is by no means exhaustive or complete, since *Uvi_Wave* contains more than 70 different functions. Detailed help can be found in the *Uvi_Wave* manual and with the MATLAB `help` command.

Filters. To use a transform filters have to be generated. The Daubechies filters of (even) length M are generated by the command

```
[h,g,rh,rg]=daub(M)
```

Here h is the low pass analysis filter, h the high pass analysis filter, and rh, rg the corresponding synthesis filters. The symlets of length M are generated with the command `symlets(M)`. Both of these families are orthogonal. The biorthogonal filters in the CDF(M,N) family are generated with the command `wspline(M,N)`. Note that *Uvi_Wave* has no function for generating Coiflets. These filters can be obtained from tables in the literature, see for example [5], or from other toolboxes, for example those mentioned in Sect. 14.2.

1D wavelet transforms. The direct and inverse transforms are obtained with the commands

```
wt(S,h,g,k)
iwt(S,rh,rg,k)
```

Here S is the signal vector to be transformed or inverted. See the documentation for the ordering of the entries in the direct transform vector. The number of scales (see Sect. 3.5) to be used is specified with k. These functions use the filter bank approach to the DWT, see Chap. 7. The transforms use periodization to deal with the boundary problem, see Sect. 10.4. Alignment is by default performed, using the first absolute maximum method, see Sect. 9.4.3. The alignment method can be changed using wtmethod.

2D wavelet transforms. The separable 2D transforms are implemented in the functions

```
wt2d(S,h,g,k)
iwt2d(S,rh,rg,k)
```

The principles are described in Chap. 6. There is a script format2d, which explains the output format in detail. Run it before using these commands.

Wavelet packet transforms. The direct and inverse wavelet packet transforms in dimensions one and two are implemented in

```
wpk(S,h,g,0,B)
iwpk(S,rh,rg,0,B)
wpk2d(S,h,g,0,B)
iwpk2d(S,rh,rg,0,B)
```

The fourth variable 0 (zero) specifies filter bank ordering of the frequencies. Change to the value 1 to get the natural frequency order. See Sect. 9.3. The basis to be used is specified in the parameter B. The basis is described according to the *Uvi_Wave* scheme, as explained in Sect. 11.8.1. Note that it is different from the one adopted in our implementations. Run the script basis for explanations and examples.

Best basis algorithm. The best basis algorithm from Sect. 8.2.2 is implemented for an additive cost function C (see Sect. 8.3) with the command

```
[basis,y,total]=pruneadd(S,h,g,C,P)
```

Here P is an optional parameter value that may be needed in the cost function, for example the value of p in the ℓ^p-norm cost function. This cost function and the Shannon entropy are implemented in the functions

```
lpenerg(S,P)
shanent(S)
```

The function pruneadd returns the selected basis in basis, the transformed signal in this basis in y, and the total cost of this representation in total. Two additional functions can be of use in interpreting the results obtained using pruneadd.

```
tfplot(B)
tree(B)
```

The first one displays the tiling of the time-frequency plane associated with basis B, as in Fig. 9.11. The second one displays the basis as a tree graph. Further explanations are given in the script basis.

Multiresolution analysis. The concept of a multiresolution analysis was explained in Chap. 4. The relevant commands are

```
y=multires(S,h,rh,g,rg,k)
split(y)
y=mres2d(S,h,rh,g,rg,k,T)
```

Here k is the number of scales. The result for a length N signal is a $(k+1) \times N$ matrix. For k less than 10 the result can be displayed using split(y). The two-dimensional version works a little differently. The output is selected with the parameter T. If the value is zero, then both the horizontal and vertical part of the separable transform used the low pass filter. Other values give the other possibilities, see the help pages. Note the ordering of the filters in these functions.

Exercises

See the exercises in Chap. 4, Chap. 5, and Chap. 6.

14. Applications and Outlook

In this chapter we give some information on applications of wavelet based transforms. We only give brief descriptions and references, since each topic requires a different background of the reader. We also give some directions for further study of the vast wavelet literature.

The World Wide Web is a good source of information on wavelets and their applications. We recommend all readers seriously interested in wavelets to subscribe to the *Wavelet Digest*. Information on how to subscribe can be found at the URL 1. The reference of the form URL 1 is to the first entry in the list at the end of this chapter.

14.1 Applications

Wavelets have been applied to a large number of problems, ranging from pure mathematics to signal processing, data compression, computer graphics, and so on. We will mention a few of them and give some pointers to the literature.

14.1.1 Data Compression

Data compression is a large area, with many different techniques being used. Early in the development of wavelet theory the methods were applied to data compression. One of the first successes was the development of the FBI fingerprint compression algorithm, referred to as Wavelet/Scalar Quantization. Further information on this particular topic can be found at the URL 2 and the URL 3.

Let us briefly describe the principles. There are three steps as shown on Fig. 14.1. The given signal s is transformed using a linear transformation T. This could be a wavelet packet transform, which is invertible (has the perfect reconstruction property). The next step is the lossy one. A quantization is performed. The floating point values produced by the wavelet transform are classified according to some scheme. For example an interval $[y_{min}, y_{max}]$ of values is selected as relevant. Transform values above and below are assigned to the chosen maximum and minimum values. The interval is divided into N subintervals of equal length (or according to some other scheme), and the interval numbers are then the quantized values. (Note that the thresholding used in Chap. 4 can be viewed as a particular quantization method.) Finally these N values are coded in order to get efficient transmission or storage of the quantized signal. The coding is usually one of the entropy coding methods, for example Huffman coding.

Fig. 14.1. Linear data compression scheme. The signal s is first transformed using a linear transform T. It is then quantized by some scheme Q. Finally the quantized signal is coded using entropy coding, to get the result s_c

The compression scheme described here is called an open-loop scheme, since there is no feedback mechanism built into it. To get efficient compression the three components T, Q, and C must be optimized as a system. One way of doing this is by a feedback mechanism, for example leading to changes in the quantization, if the coding step leads to a poor result, by some measure. The overall design goal is often that the compressed and then reconstructed signal should be sufficiently close to the original, by some measure. For example, the average person listening to music being played back from a compressed version should not notice any compression effects, or the compression effects should be acceptable, by some subjective measure.

The type of compression used, and the level of compression that is acceptable, depends very much of the type of application one considers. For example, in transmission of speech the first requirement is intelligibility. The next step is that the speaker should be recognizable, etc. For music the requirements are different, and more difficult to satisfy. One often uses a model based approach. Statistical methods are also often used.

In image compression the issues are even more complex, and the problems are compounded by the fact that it is difficult to find good models for images. A recent development is the use of separable 2D wavelet transforms, defined using lifting, in the new image compression standard JPEG2000. The previous standard JPEG was based on block discrete cosine transforms. See URL 4 for further information.

Continuing with video compression and multimedia applications, we get to the point where the results are current research. We should mention that parts of the MPEG-4 standard will use wavelet based methods. Try looking at the URL 5, or search the Web for sites with information on MPEG-4.

Let us also note that the data compression problem has an interesting mathematical aspect, in the problem of characterizing classes of signals, in particular images, and based on the classification, trying to devise efficient compression schemes for classes of signals.

It is clear from this very short description that there are many aspects beyond the wavelet transform step in getting a good compression scheme. We refer to the books [16, 28] for further results, related to wavelet based methods.

14.1.2 Signal Analysis and Processing

One of the applications shown in Chap. 4 was to signal denoising, see in particular Fig. 4.11. Our approach there was to select a threshold by visual inspection of a number of trials. See also the examples in Sect. 13.3. To be used in practice one needs a theory to determine how to select a threshold, or some other method. Several results exist in this area. We refer to the book [16] and the references therein. Again, this is an active area of research, and many further results are to be expected.

Another area of application is to feature identification in signals. For one-dimensional signals (for example seismic signals) there is considerable work on the identification of features in such signals. In particular, singularities can be located very precisely, as shown in simple examples in Chap. 4. The edges in a picture can also be detected, as shown in Fig. 6.9, but this is actually a rather complicated issue. Identification of other structures in images is again an area of current research.

We should also mention that for one-dimensional signals the time-frequency planes discussed in Chap. 9 can be a good starting point for the analysis of a class of signals. Again, much more can be done than was described in that chapter. In particular, the analysis can also be performed using transforms based on the discrete cosine transform.

14.1.3 Other Applications

Applications to many other areas of science exist. For some examples from applied mathematics, see the collection of articles [14], and also the book [18]. For applications to the solution of partial differential equations, see [9]. The collection of articles [1] contains information on applications in physics. The books [21] and [29] discusses applications in statistics. Applications to meteorology are explained at the URL 6. Many other applications could be mentioned. We finish by mentioning an application involving control theory, in which one of the authors is involved, see the URL 7.

14.2 Outlook

First a word of warning. Application of wavelets involves numerical computations. There are many issues in computational aspects that have not been touched upon in this book. On thing should be mentioned, namely that application of the lifting technique can lead to DWTs that are numerically unstable. Our suggestion is at least initially to rely on the well known orthogonal or biorthogonal families, where it is known that these problems do not occur. If the need arises to construct new transforms using lifting, one should be aware of the possibility of numerical instability.

The reader having read this far and wanting to learn more about wavelets is faced with the problem of choice of direction. One can decide to go into the mathematical aspects, or one can learn more about one of the many applications, some of which were mentioned above. There is a large number of books that one can read. But one should be aware that each book mentioned below has specific prerequisites, which vary widely.

Concerning the mathematical aspects, then the book by I. Daubechies [5] might be a good starting point. Another book dealing mainly with mathematical aspects is the one by E. Hernández and G. Weiss [12]. For those with the necessary mathematical foundations the book by Y. Meyer [17], and the one by Y. Meyer and R. Coifman [19], together provide a large amount of information on mathematical aspects of wavelet theory. There are many other books dealing with mathematical aspects. We must refer the reader to for example Mathematical Review, were many of these books have been reviewed.

A book which covers both the mathematical aspects and the applications is the one by S. Mallat [16] that we have mentioned several times. It is a good source of information and pointers to recent research. There are many other books dealing with the wavelets from signal analysis point of view. We have already referred to the one by M. Vetterli and J. Kovačević [28]. It contains a lot of information and many references. The book by G. Strang and T. Nguyen [24] emphasizes filters and the linear algebra point of view.

The understanding of wavelets is enhanced by computer experiments. We have encouraged the reader to do many, and we hope the reader has carried out all suggested experiments. We have based our presentation on the public domain MATLAB toolbox *Uvi_Wave*, written by a number of scientist at Universidad de Vigo in Spain. It is available at the URL 8. Here one can also find a good manual for the toolbox. There are some problems in using this toolbox with newer versions of MATLAB. Their resolution is explained at the URL 11.

For further work on the computer there exists a number of other toolboxes. We will mention two of them. One is the public domain MATLAB toolbox WaveLab. It contains many more functions than *Uvi_Wave*, and it has been updated to work with version 5.3 of MATLAB. The many possibilities also mean that it is more demanding to use. Many examples are included with the toolbox. WaveLab is available at the URL 9. The other toolbox is the official MATLAB Wavelet Toolbox, which recently has been released in a new version. It offers a graphical interface for performing many kinds of wavelet analysis. Further information can be found at the URL 10. There exist many other collections of MATLAB functions for wavelet analysis, and libraries of for example C code for specific applications. Again, a search of the Web will yield much information.

Finally, we should mention once more that the M-files used in this book are available at the URL 11. The available files include those needed in the implementation examples in Chap. 11, and all examples in Chap. 13. There is also some additional MATLAB and C software available, together with a collection of links to relevant material. It is also possible to submit comments on the book to the authors at this site.

14.3 Some Web Sites

Here we have collected the Web sites mentioned in this chapter. The reader is probably aware that the information in this list may be out of date by the time it is read. In any case, it is a good idea to use one of the search engines to try to find related information.

1. http://www.wavelet.org
2. http://www.c3.lanl.gov/~brislawn/FBI/FBI.html
3. ftp://wwwc3.lanl.gov/pub/misc/WSQ/FBI_WSQ_FAQ
4. http://www.jpeg.org
5. http://www.cselt.it/mpeg/
6. http://paos.colorado.edu/research/wavelets/
7. http://www.beamcontrol.com
8. ftp://ftp.tsc.uvigo.es/pub/Uvi_Wave/matlab
9. http://www-stat.stanford.edu/~wavelab/
10. http://www.mathworks.com
11. http://www.bigfoot.com/~alch/ripples.html

References

1. J. C. van den Berg (ed.), *Wavelets in physics*, Cambridge University Press, Cambridge, 1999.
2. A. Cohen, I. Daubechies, and J.-C. Feauveau, *Biorthogonal bases of compactly supported wavelets*, Comm. Pure Appl. Math. **45** (1992), no. 5, 485–560.
3. A. Cohen, I. Daubechies, and P. Vial, *Wavelets on the interval and fast wavelet transforms*, Appl. Comput. Harmon. Anal. **1** (1993), no. 1, 54–81.
4. A. Cohen and R. D. Ryan, *Wavelets and multiscale signal processing*, Chapman & Hall, London, 1995.
5. I. Daubechies, *Ten lectures on wavelets*, Society for Industrial and Applied Mathematics (SIAM), Philadelphia, PA, 1992.
6. _____, *Orthonormal bases of compactly supported wavelets. II. Variations on a theme*, SIAM J. Math. Anal. **24** (1993), no. 2, 499–519.
7. I. Daubechies and W. Sweldens, *Factoring wavelet transforms into lifting steps*, J. Fourier Anal. Appl. **4** (1998), no. 3, 245–267.
8. J. H. Davenport, Y. Siret, and E. Tournier, *Computer algebra*, second ed., Academic Press Ltd., London, 1993.
9. S. Goedecker, *Wavelets and their application*, Presses Polytechniques et Universitaires Romandes, Lausanne, 1998.
10. C. Herley, J. Kovačević, K. Ranchandran, and M. Vetterli, *Tilings of the time-frequency plane: Construction of arbitrary orthogonal bases and fast tiling algorithms*, IEEE Trans. Signal Proc. **41** (1993), no. 12, 2536–2556.
11. C. Herley and M. Vetterli, *Wavelets and recursive filter banks*, IEEE Trans. Signal Proc. **41** (1993), no. 8, 2536–2556.
12. E. Hernández and G. Weiss, *A first course on wavelets*, CRC Press, Boca Raton, FL, 1996.
13. B. Burke Hubbard, *The world according to wavelets*, second ed., A K Peters Ltd., Wellesley, MA, 1998.
14. M. Kobayashi (ed.), *Wavelets and their applications*, Society for Industrial and Applied Mathematics (SIAM), Philadelphia, PA, 1998.
15. W. M. Lawton, *Necessary and sufficient conditions for constructing orthonormal wavelet bases*, J. Math. Phys. **32** (1991), no. 1, 57–61.
16. S. Mallat, *A wavelet tour of signal processing*, Academic Press Inc., San Diego, CA, 1998.
17. Y. Meyer, *Wavelets and operators*, Cambridge University Press, Cambridge, 1992.
18. _____, *Wavelets, algorithms and applications*, SIAM, Philadelphia, Pennsylvania, 1993.
19. Y. Meyer and R. Coifman, *Wavelets*, Cambridge University Press, Cambridge, 1997.
20. C. Mulcahy, *Plotting and scheming with wavelets*, Mathematics Magazine **69** (1996), no. 5, 323–343.

21. P. Müller and B. Vidakovic (eds.), *Bayesian inference in wavelet-based models*, Springer-Verlag, New York, 1999.
22. A. V. Oppenheimer and R. Schafer, *Digital signal processing*, Prentice Hall Inc., Upper Saddle River, NJ, 1975.
23. A. V. Oppenheimer, R. Schafer, and J. R. Buck, *Discrete-time signal processing*, second ed., Prentice Hall Inc., Upper Saddle River, NJ, 1999.
24. G. Strang and T. Nguyen, *Wavelets and filter banks*, Wellesley-Cambridge Press, Wellesley, Massachusetts, 1996.
25. W. Sweldens, *The lifting scheme: A custom-design construction of biorthogonal wavelets*, Appl. Comput. Harmon. Anal. **3** (1996), no. 2, 186–200.
26. _____, *The lifting scheme: A construction of second generation wavelets*, SIAM J. Math. Anal. **29** (1997), no. 2, 511–546.
27. G. Uytterhoeven, D. Roose, and A. Bultheel, *Wavelet transforms using the lifting scheme*, Report ITA-Wavelets-WP1.1 (Revised version), Department of Computer Science, K. U. Leuven, Heverlee, Belgium, April 1997.
28. M. Vetterli and J. Kovačević, *Wavelets and subband coding*, Prentice Hall Inc., Upper Saddle River, NJ, 1995.
29. B. Vidakovic, *Statistical modeling by wavelets*, John Wiley & Sons Inc., New York, 1999.
30. M. V. Wickerhauser, *Adapted wavelet analysis from theory to software*, A K Peters Ltd., Wellesley, MA, 1994.

Index

Index